Modifying Our Genes

Modifying our Genes

Theology, Science and 'Playing God'

Alexander Massmann
and
Keith R. Fox

scm press

© Alexander Massmann and Keith Fox 2021

Published in 2021 by SCM Press
Editorial office
3rd Floor, Invicta House,
108–114 Golden Lane,
London EC1Y 0TG, UK
www.scmpress.co.uk

SCM Press is an imprint of Hymns Ancient & Modern Ltd
(a registered charity)

H
Y Ancient
M &Modern
N
S

Hymns Ancient & Modern® is a registered trademark of
Hymns Ancient & Modern Ltd
13A Hellesdon Park Road, Norwich,
Norfolk NR6 5DR, UK

Scripture quotations are from New Revised Standard Version
Bible: Anglicized Edition, copyright © 1989, 1995 National
Council of the Churches of Christ in the United States of America.
Used by permission. All rights reserved worldwide.
And from the Holy Bible, New International Version. Copyright
© 1973, 1978, 1984, 2011 by Biblica, Inc. Used by permission of
Zondervan. All rights to be reserved worldwide.

The Authors have asserted their right under the Copyright, Designs
and Patents Act 1988 to be identified as the Authors of this Work

British Library Cataloguing in Publication data

A catalogue record for this book is available
from the British Library

978-0-334-05953-0

Typeset by Regent Typesetting
Printed and bound by
CPI Group (UK) Ltd

Contents

v

Acknowledgements

This book on the ethics of genome editing in humans has grown out of a research project at the Faraday Institute for Science and Religion (Cambridge, UK). We are grateful to its 'Uses and Abuses of Biology' grants programme, which was funded by the Templeton World Charity Foundation, for financial support. We also thank The Faraday Institute and the various people with whom we have worked together there. Among them, Dr Denis Alexander has been a helpful partner in dialogue. We would also like to thank our editor David Shervington.

A. M. and K. R. F., October 2020

Introduction

In November 2018, a Chinese scientist made headline news across the world. At an otherwise routine meeting of specialists, the generally little-known Dr He made an unprecedented announcement: he had modified the DNA of human embryos, which he then implanted in two women, leading to successful pregnancies and the birth of two babies. The technique involved is called genome editing, which makes use of a relatively new piece of technology known as CRISPR-Cas9. The goal of the procedure was to make the children resistant to infection by HIV. Although the results have not been properly scrutinized and published, the procedure raised some profound ethical and legal questions, for which many people were unprepared. Researchers have quickly taken up CRISPR-Cas9 as an experimental tool, which they use on cells grown in the laboratory. However, many scientists had already called for a moratorium on the use of this tool on human embryos. Some of their concerns were about issues of safety and risk, though the genetically modified babies raised many other questions about whether we should be tinkering with human genetics. In many countries across the world, bringing about a pregnancy with a modified embryo is illegal. However, some people see this as the start of a bright new future for the use of genome editing in humans. What should we think about such procedures, and should laws be changed?

What is genome editing?

At the heart of the debate is CRISPR-Cas9, or CRISPR for short (pronounced *'crisper'* – and for the record, the acronym stands for 'clustered regularly interspaced short palindromic repeats'). In 2012, an international research team discovered how to engineer a bacterial enzyme, which then allowed them to cut and modify DNA in a very precise way. DNA is the molecule that stores the genetic information in the cells of every living being. The reason for the excitement around CRISPR-Cas9 is that this enzyme works like a scalpel for genetic surgery, cutting DNA at any chosen location, with a precision that was previously unimaginable. This is the molecular equivalent of the 'find and replace' function in a word processing programme – hence the nickname 'genome editing'. In 2020, the Nobel Prize in Chemistry was awarded to two researchers for their work in developing CRISPR, Jennifer Doudna and Emmanuelle Charpentier. It has now become widely used as a powerful research tool for studying how cells work. Researchers appreciate that it is relatively inexpensive and easy to use. The modification of human embryos in human reproduction marks a new stage, however.

Genetic engineering has been with us since the 1970s, for example allowing us to produce insulin for treating diabetes and to make plants more resistant to drought and disease. However, the new genetic editing tool, CRISPR, allows researchers to rearrange the genetic material with an ease and precision that makes traditional genetic engineering look clumsy and time consuming. It has great potential for treating or preventing several genetically inherited diseases. With the enhancement of capabilities in people who are not even ill, it might in principle allow for much more ambitious projects, perhaps for increasing muscle strength, a more refined sense perception or a more powerful memory. The application of CRISPR-Cas9 to human biology has brought important ethical questions to the fore, and that is the topic we are discussing in this book.

The excitement about CRISPR has already spilled into pop culture, which has focused on the possibility of its misuse and abuse. The technology has inspired an episode of the TV series *The X-Files*, and in 2016 Jennifer Lopez was reported to have signed to produce a TV series in which CRISPR is put to use for dark machinations. Seventy years after Aldous Huxley's

Brave New World, genetic modification has inspired further dystopian novels like Margaret Atwood's *Oryx and Crake* (2003) and Peter James's *Perfect People* (2011).[1] Even before the discovery of CRISPR, the Hollywood drama *Gattaca* (1997) involved genetically modified humans. For the title of the film, the makers played with the four letters (G, A, T and C) that are commonly used to abbreviate the four key building blocks of DNA. These writers and film makers warn that, whether we're looking for a quick medical fix or to take human capabilities to new levels, we may actually produce other social and personal issues that ultimately reduce our humanity.

In contrast, proponents of far-reaching genetic interventions are weary of what they consider to be scaremongering. They suggest that interventions into the genome have the potential to bring about some significant positive changes. If that is so, then it appears that the pessimism of the critics, not the technology, is the problem.

Nevertheless, most researchers, as well as the general public, were outraged at the modification of babies in China, arguing that while CRISPR has extraordinary potential, the technology is not yet ready for use in humans. At the moment, despite its precision, CRISPR may cause some unintended changes in DNA (so-called 'off-target events') and so could seriously harm health and well-being. At the very least, the technology needs to be refined before it should be used in humans. Until then, it's certainly not safe for work on human embryos, and its use could produce harmful and irreversible effects. However, we should not overstate these reservations. Some people are against using CRISPR in human embryos in principle (and some of these are against all procedures that involve human embryos), though many others accept the possibility of this type of genetic modification once the technology can be applied safely. For many, it is just that a highly promising technology should not be compromised by a premature roll-out. These people anticipate that, once the technology has been optimized, the laws will be changed to allow for the generation of genetically modified babies.

Even if the technology can be made to work without any

off-target effects, it may still be too risky. This is because genes do not act in isolation, as if each had one clear-cut, direct effect on the body. Over the past decades, researchers have used their detective skills to unravel the complex networks in which one gene interacts with other genes. Modifying the activity of one gene could have unintended consequences on other genes in the network, tipping a delicate balance and producing unexpected changes. The recent example of the Chinese researcher who modified the embryos of two babies illustrates the point. His stated goal was to make the children immune to HIV, but it has been argued that his particular intervention – knocking out the CCR5 receptor gene – may not only ensure resistance to HIV, but also increase the risk of infection by the West Nile virus or influenza. The human body is fascinating, but it is also immensely complicated.

While some people see science as making great strides, for others the risks are too serious to contemplate. As we venture into uncharted territory, are we opening the door for its misuse? Some risks can be carefully evaluated, but the challenges of the technology are far more complicated than simply optimizing risks and benefits. Given the complexity of the project, it is also important to ask whether we have exhausted all other possible means to alleviate or eliminate serious disease before we turn to genome editing. For example, many inherited genetic conditions can be prevented by pre-implantation genetic diagnosis, in which embryos are screened to see which ones are 'healthy' and which ones contain a defective gene.

Many people still feel uneasy about producing genetically modified babies, and simply balancing the risks and benefits will not deal with the whole spectrum of religious and moral questions. Some think that this is 'playing God' and that we are venturing into places where humans should not go.

Is it possible that scientists might overstep a boundary and reach too deeply into the fabric of nature, regardless of the consequences? What would the modification of an embryo mean for people who believe that nature is God's creation and that creation is good? Others suggest that we should not 'meddle with Mother Nature'. Admittedly, that's a rather vague view of how nature and science work, but is how many people think about complicated and specialized scientific work. Even the cutting-edge scientist Jennifer Doudna, who, as noted, was awarded a Nobel Prize for her part in developing CRISPR-Cas9, used religious language, although she is not known as a religious person. Musing about a potential catalogue of desirable traits from which parents may choose for their children's genetic improvement, she struck a critical note, saying that it's 'incredibly important' to have discussions about 'how to protect the sanctity of human life'.[2]

Although the natural world is very special, we know that it is often not benign and unambiguously good. Christians, Muslims and Jews affirm that the world is God's good creation, but in our experience, it often does not appear to work that way. And so, for centuries, medical practitioners have sought to treat diseases and to alleviate some of the pain and suffering that is inherent in nature. In this regard, the eradication of pathogens like smallpox was a drastic intervention, though most people would surely consider this to be a positive step. Wiping out the bacteria that cause tuberculosis, cholera or the bubonic plague would also be a good thing. So endeavours to improve life that interfere with nature should not always be seen as negative. The mere fact that scientists modify DNA, rather than something else, cannot be a sufficient criterion for deciding between good or bad interventions. When the twentieth-century theologian Karl Barth addressed health and illness as part of his wider discussion of God's work in creation, he emphasized that Christians must resolve to pursue health and not to accept disease too lightly, with a misplaced faith that legitimizes all reality as the will of the creator. Rather, he asserted that the creator intended a full measure of life for each creature.[3] And so medical treatments and care for the sick have been a distinctive part of Christian action since the very early centuries.

There should be little doubt that *some* forms of genetic modifications in humans could be beneficial. CRISPR has given scientists high hopes that we can have a major impact on the 10,000 or more heritable diseases, many of them severe, that are caused by specific genetic changes and which are otherwise incurable. Might this technology also be used to produce an effective cure for cancer? In this context some applications of genome editing in humans have been largely uncontroversial. For example, in 2015 a London hospital successfully treated two young children with a genome-editing technology closely related to CRISPR-Cas9. They had been suffering from a severe form of leukaemia that would almost certainly have killed them. The physicians introduced genetically modified cells into the bodies of the two girls when they were about one year old. There have also been similar promising clinical tests for a cure for haemophilia and sickle-cell disease, and a lot of money has already been invested to make beneficial genome-editing solutions available to doctors and patients.

The most pressing question, however, is whether to allow the genetic modification of a human embryo in order to implant it, so that the mother would carry the child to term. This is what the Chinese researcher did – though he is now serving a three-year prison sentence with a large fine. Many countries have laws that forbid pregnancies with genetically modified embryos. However, if the tools for modifying embryos can be refined, then this would provide a much more powerful procedure than modifying the genes of children or adults. For example, if a child already has cystic fibrosis, then genome editing would need to modify a lot of cells to initiate a cure, and this may be technically unrealistic. However, by modifying the few cells of an embryo, this heritable disease would be completely removed from the individual, and the condition would not be passed on to further generations. Many people see this as a persuasive reason for changing the existing laws on genetically modifying embryos.

A survey in the UK suggested that a significant part of the British population would cautiously embrace the genetic modification of embryos.[4] While respondents rejected the suggestion that parents could edit genes to choose their baby's hair colour, they generally agreed that we could use CRISPR in embryos to combat disease, and they even warmed to the idea that one might use it to raise a child's IQ score – if that is ever a realistic possibility. Professional bodies too have agreed with the use of genome editing for medical purposes. In an extensive report in 2017, the American Academies of Sciences and of Medicine argued that the genetic modification of embryos should be legal if it prevents or alleviates disease, a stance they largely reiterated in 2020. The UK Nuffield Council on Bioethics largely agreed in 2018, and even supported certain non-medical genetic 'improvements'.[5] Taken together, all these developments have set the stage for a wider public debate about the question of whether the UK should legalize the genetic modification of human embryos, and if so, for what purposes.

What we will do in this book

This book will first help you understand how CRISPR works and what can be done with this technology. While there are many fascinating applications of the technology, such as modifying mosquitoes in order to wipe out malaria, we will focus on targeted genetic modifications in humans. Within the broad topic of genetic technologies that can be used in humans, we will not discuss topics such as direct-to-consumer services for testing genetic predispositions or techniques for the genetic testing of a fetus in the womb. Neither will we directly address controversial procedures such as cloning or mito-chondrial replacement therapy. These are important topics in their own right, but this book focuses on the new moral questions that are raised specifically by genome editing. In theory, CRISPR allows us to modify specific genes in targeted ways, while leaving the rest of the genome unchanged. The biotechnological possibilities are truly remarkable, though some of the suggestions may not be realistic or may be a long time off. It is therefore important to have a good sense of the technical possibilities before asking which uses of the new technology we should encourage or discourage.

Our own views on the ethics of genome editing are those of a Christian theologian and ethicist (Alexander Massmann) and of a scientist and professor of biochemistry, who is also a committed Christian (Keith Fox). We are aware that we live in a pluralist society, together with people of other faiths and none, who may come to similar or different conclusions. In many parts of the world, including the UK, continental Europe and the USA, producing a pregnancy with a genetically modified embryo is illegal. However, some people hope that the law will change, and we expect a larger political debate on genome editing to take place. In that debate, Christians cannot assume that their views will be privileged over others, and we will try to develop coherent ethical arguments that make sense to non-Christians as well, without the need to quote heavily from Bible passages. To argue a moral case from a Christian perspective does not mean that we simply state whether God wants things done one way or another; our concern is rather to ask what is at stake, what would be best for individuals and for the common good, to contribute to a tolerant and inclusive society – within a Christian framework, but also with continuous focus on the question of where we could find common ground with citizens who do not share our religious commitments.

One cannot write about a public policy question, such as genome editing, simply by addressing Christians, as if they are the only ones who will be affected. In fact, we learn many helpful things in dialogue with those who hold different world views. This is an important part of how we think about moral questions, and we hope that a genuine discussion will be possible. Christians will of course want to remain faithful to their central beliefs, but as one Christian ethicist put it, Christians should respect the beliefs of others 'sufficiently to strive to persuade them; and that successful persuasion involves the imaginative inhabiting of the other's point of view'.[6]

In our willingness to learn, it is important to us also to bring the rich heritage of biblical traditions to the table, which can shed light on moral questions. For example, Jesus' healing ministry reminds us of the importance of physical integrity – a heritage that has contributed significantly to wider Western culture. At the same time, in the writings of the Apostle Paul, we see the will to carve out a meaningful life when faced with abiding physical challenges. We are well aware that Christians often do not agree with each other on moral issues. Nonetheless, that should not hinder us from formulating our own arguments and from trying to understand where other people are coming from. Even if the thinkers with whom we're in dialogue or the readers may not agree with all of our conclusions, it is important that we understand and respect our differences and learn as much as we can along the way.

We will strive to make moral sense of scientific and technological advances by drawing not only on resources from the Judaeo-Christian tradition but also on fields as diverse as genetics, philosophy, history and psychology. Christian ethics is always done best in an interaction that also involves non-theological resources, in order to make sense of reality in its bewildering complexity.

Getting our heads around the science

To begin with, it is important to understand what CRISPR is and what it can do, especially as there can be confusion about what it means when we talk about modifying 'a person's genes'. Although we each have a unique genome – the complete collection of our genes – these are copied and stored in each of the trillions of cells that make a complete human body. So where should we start …?

In this book, we will first equip readers with a better grasp of what genome editing is, breaking down the science in layperson's terms. What exactly is genome editing and why is it now so much easier to use than the previous types of genetic engineering? How and why could this be used on human embryos, and how does this differ from using it to affect fully formed humans (children and adults)? Some of the potential applications may seem like far-fetched science fiction, but the science is progressing rapidly and today's science fiction could become tomorrow's science fact. So we need to think about developments that are presently hard to predict. At the same time, that does not mean that the most fantastical scenarios carry the same weight as procedures that have already been successfully tested in clinical trials. At any rate, the science is fascinating and it will lay the groundwork for further moral discussions.

It is important to get this right. Many aspects of the science of genetics are well established, but it is easy to get confused, especially when media hype overstates or oversimplifies the importance of genes. One journalist, describing the four key chemical building blocks of DNA, states: 'All the instructions for building and running the human body are encoded in combinations of those four [chemical] bases.' Genes are important, often crucially important, but to call them 'all the instructions' for the operation of the body is rather an overstatement. Consider the effort required for a toddler's first steps on two feet. Of course, our genes support the processes that allow humans to walk on two legs rather than four, but genes alone are not enough to send a toddler wobbling across the room! The way that we think about genes will strongly influence how we think about genome editing: do we overestimate or underestimate the contribution that genes make to our lives? The more we emphasize the importance of genes, the more polarized the debate about genome editing becomes: for some it appears that there are few problems that genome editing cannot solve, while for others, the warning that we must not 'play God' becomes ever more urgent.

Another example of the oversimplification of the importance of genes can be seen in recent news items. Many journalists reported on the work of George Church, a highly regarded geneticist who hopes to use genome editing to bring the woolly mammoth back from extinction. The impression is given that, once the right DNA changes are in place, and given the right equipment and some time, a mammoth will be produced. However, the resulting animal will have no knowledge of how to forage in icy conditions or avoid getting bogged down in a swamp. The fact is that behaviours like that are learned, not biologically inherited. In Africa, for example, elephants pass knowledge on from generation to generation. The fate of an entire elephant herd may depend on the old matriarch who remembers one particular watering hole from decades ago.

In the case of the mammoth, this long chain of learning was permanently broken when the last of these behemoths died thousands of years ago, and no amount of genome editing will make this knowledge available to modern-day mammoths, even if they can be physically resurrected. If the role of genes can be overestimated in mammoths, we should be especially wary of a similar error in human applications.

Although it is easy to overestimate the power of genes, we also know that very small genetic changes can have dramatic effects and produce life-changing medical conditions. This is especially true for the thousand or more genetic diseases that result from very small changes in DNA sequence – sometimes as little as one simple change among the three billion letters in the human genome. Many of these conditions are extremely rare, but some are more common. For example, sickle-cell disease is one of the most common inherited diseases, and it affects about 13,000 people in the UK and about 100,000 Americans.[7] Similarly, cystic fibrosis affects at least 30,000 people in the USA and about 10,000 people in the UK. It is most common in Caucasians of Northern European ancestry. While not having the disease themselves, about one in 29 people of Caucasian ancestry can pass on the mutation of the cystic fibrosis gene to the next generation. So we see that a tiny genetic change can have a serious impact on a person's life. Some people's conditions can be accommodated by altering diet or modifying lifestyle, but for others life becomes a struggle or is cut short. Interest in medical interventions for such conditions has led to intense research into the possibility of using CRISPR as a means of correcting the underlying errors. However, the causes of many conditions are much more complicated and arise from the combined action of many genes. For example, autism is influenced by several genetic factors working together that combine with environmental influences to produce a spectrum of effects. Physical characteristics such as height and intelligence are also affected by combinations of genes, each of which alone only has a very small effect.

Another problem is that there can be confusion about what people mean by the phrase 'genetic modification'. A survey of public opinion in the UK, done for the Royal Society, asked respondents whether they would approve of 'heritable' genetic changes, without mentioning that such powerful changes would have to be done on an embryo, even before it starts to develop in the mother's womb. This would require in vitro fertilization (IVF) instead of natural intercourse. While the burdensome procedure itself could be an obstacle for some, the genetic modification of embryos and their subsequent implantation in the mother's womb is a very sensitive issue, especially for many religious people. Certain embryos are likely to be destroyed or discarded in the process, and for Roman Catholics and some other Christians, the destruction of embryos determines their opposition to the genetic modification of them as a matter of principle. However, not all Christians take this stance, and there is diversity of opinion on the status of an embryo that does not necessarily align with more traditional or more liberal theologies. Another crucial point is that some existing alternatives (such as pre-implantation genetic diagnosis) would make many genetic modification of embryos unnecessary. This aspect is often not mentioned when the issues are discussed, and these technical points can dramatically affect the discussions on whether we should legalize genetic modifications of embryos. As fellow Christians, we believe that a better understanding of the scientific information should help to defuse the controversy. At the very least, public surveys need to communicate more clearly to their lay respondents what they are talking about, or risk influencing the debates with well-intentioned hot air.

With lots of talk about changes to embryos, the genetic modifications that are done on children or adults have received less coverage. Such therapeutic genetic modifications are ethically less controversial, but can often have highly significant medical outcomes.

Finally, what do we make of claims that genetic interventions could be used to increase a child's intelligence or give it greater athletic prowess? Rather than healing a medical condition, this would be a form of genetic 'enhancement'. Here, too, things can look different with a better sense of the technical details: what works, how and under what circumstances? Even respected scientists come up with suggestions that are difficult to take seriously: physical modifications for easier space travel, for instance! However, even in this area, some of the things that seem like science fiction might quickly become science fact. Other suggestions that some people might find appealing, such as a genetic improvement of one's memory, may also sound far-fetched, but should not be so quickly dismissed as unrealistic.

Entering the moral debate

Once we have tackled the most important part of the science, we will ask not only what CRISPR *can* or *might* be used for but what it *should* be used for. That is the question of ethics: what reasons are there to act in one way rather than another, to employ CRISPR in this way or that? Under which circumstances and why? Should things be left to personal preference and individual choice? These are big questions that people have started to address, though to date they have largely been limited to specialized academic publications.

Following a chapter explaining the science behind CRISPR, Chapter 3 will discuss medical uses of genome editing. Should we use CRISPR to embark on a crusade to eradicate all disease? People with disabilities have repeatedly been told that a 'cure' for their particular impairments is just around the corner. If we concentrate on the medical possibilities of genome editing, are we also going to treat disabilities as if they were merely medical problems to be fixed? Would we then fail to respect how people with disabilities identify with their conditions, and regard them as alternative and equally valid ways of living? Theologians have long argued about a remark by the Apostle Paul that 'thorn was given to me in the flesh' – presumably a reference to a debilitating ailment – but this has something important to contribute here, as he saw that God's 'power is made perfect in weakness' (2 Cor. 12.7, 9). What seems like disease and weakness to some can in fact include significant strengths and opportunities for others. Here, we all need to learn from people with 'disability', and not simply to think of these as weaknesses. For example, the activist Raúl Aguayo-Krauthausen, who has brittle bone disease and uses a wheelchair, demonstrated his resilience and humour when he entitled his autobiography, *I didn't want to become a roofer anyway!*[8] On the other hand, which expectant mother would not seek to improve the health of her baby by healthy eating or avoiding alcohol and tobacco? In the end, we have to ask what the significance of medical treatments is for those who suffer from disease – and how does disability differ from illness?

When a company working for the Royal Society – the UK's academy of sciences – held workshops to start wider discussions about genome editing in the UK, the result of one meeting was a poster proclaiming the 'end of all disease'. Many media reports take a similar line.[9] As tantalizing as that may sound, CRISPR is not even going to eradicate all *heritable*, or genetic, diseases, let alone infectious diseases, mental illness, and impairments due to accidents or noxious environmental influences. Efforts to produce new methods may eventually help a significant number of people but, in such a worthy-sounding endeavour, many will also be left behind, and they will contrast with the ideal if the ambition is to eradicate disease.

Here part of the ethical task is to take a closer look at the convictions and values that shape the present social situation. When we talk about eradicating disease, it is important that the people who still suffer from the condition should not be forgotten or overlooked. Of course, people should expect very good medical care, but all people should also have a right to be fully included in society, regardless of their impairments or illnesses. Several religious traditions have preserved a sense that 'the least of these' (Matt. 25.40, 45) should expect society to support them. This will require much more than simply considering medical progress. The investment that will be made into medicine will need to go hand in hand with increased attention to those who might otherwise be left behind. An overblown enthusiasm for genome editing runs the risk of excluding people whose impairment cannot simply be edited away. If an emphasis on medical perfection produces greater social marginalization of some, then we are left wondering whether the gains in medical treatment will really make life better for the everyone. What about genome editing for those who are healthy? Should we even consider the possibility of 'improving' the capabilities of healthy people, for example by raising intelligence through a genetic modification? This is the question of Chapter 4. The idea of enhancements produces an immediate negative reaction for many people, though others ask why genetic 'improvements' should be forbidden, especially when we already use non-medical 'enhancements' such as private academic tutoring. Opinions on this topic diverge strongly. One critic asks, 'What will happen to political rights once we are able to, in effect, breed some people with saddles on their backs, and others with spurs?'[10] At the other extreme are arguments about personal liberty, and one advocate of allowing genome enhancements responded by claiming that we are living in a 'moral Stalinism in which authorities dictate what we can do and what we can't'.[11]

While such radical oversimplifications, on either side, make gripping headlines, they will not help us reach a balanced moral conclusion. However, there are careful thinkers on both sides, including some who wonder why we should not allow moderate 'enhancements'. One example could come from a recent report on scientific developments arguing that genetic analysis might 'predict life expectancy'. The ultimate motive may be to 'tweak' longevity, even though this does not target any illness in the conventional sense. Should we really be so strict that we categorically forbid such genetic modifications?

Before we can comment on the rights and wrongs of non-medical enhancements, we will need to be able to define exactly what we mean by health and illness. Health seems to be a word that everyone uses, yet no one knows its full meaning! Without a clear distinction, it will not be possible to affirm what are medical uses of genome editing and what are genetic enhancements. This is one of the reasons why the UK Nuffield Council on Bioethics argued that we should allow some interventions that are otherwise regarded as non-medical enhancements. We will argue that we first need to clarify a few basic ideas in order to arrive at a useful understanding of health and illness, which will help us distinguish between medicine and enhancement.

However, clarifying the distinction between therapy and enhancements will not, on its own, answer the question of whether we should allow genetic enhancements. Most parents feel responsible for their child's well-being, but if parents genetically 'improve' their baby, would we end up with 'premium kids'? Might enhancements increase the trend towards 'helicopter parenting'? Or would so-called genetic enhancements increase life satisfaction, as proponents claim? Perhaps the worries are not as straightforward as they seem – no genetic fix will automatically result in stronger muscles; the hard work of training and sweating will still be necessary to achieve greater fitness. An athletic enhancement, if such a thing is possible, would increase the *potential* for fitness, but would not make it an immediate reality. So the personal choice might still belong to the child, albeit under the influence of the parents' expectations. Would that be enough to satisfy the critics?

The suggestion that greater physical or mental abilities could lead to a better and more fulfilled life has some intuitive appeal. The ancient theologian Irenaeus of Lyon is sometimes quoted as saying that 'the glory of God is the human person fully alive'.[12] If that is so, would not a more powerful human body, living life more fully, be better from a faith perspective too? Others argue that if humans are 'created in the image of God', then we might be called to take practical steps in an ongoing, positive form of cooperation with God the creator.[13]

In increasing capabilities through enhancements, some suggest that we should aim to prevent disability, while still caring for the disabled.[14] These people hope to reclaim a moral core of a 'liberal eugenics', with an emphasis on free decision-making. This, in Chapter 5, brings the important issue of eugenics up for discussion. Opponents of 'a 'liberal eugenics', by contrast, argue that eugenics is irredeemably beyond the pale. When a contemporary science writer contemplates whether we should 'breed humans for mathematical, musical or athletic ability',[15] already the language is degrading, even if he does not answer his question with a *yes*. It may not require a Nazi world view for eugenics to be morally wrong. Indeed, there is a wider history of eugenics, apart from Nazi practices, that can inform our discussion of genome editing. We will also explain why many eugenic suggestions are typically bad science. Finally, Chapter 6 discusses more broadly what it means to be human in an age of biotechnology. What larger views shape the way that society should treat the human body? One of the crucial questions is how we evaluate higher physical functioning. Is it just a higher point on a sliding scale of life quality, with standard health ranking a little further down, just above disease? Can we evaluate physical limitations and possibilities and hope that genome editing will help us shape our bodies accordingly? Are there better ways that will help us to evaluate how particular physical traits contribute to a meaningful life? What role do the voices of people with disabilities play in the conversation? And how do religious views or moral traditions contribute to those hopes and fears?

In a recent contribution to this debate, one theologian argued that God has already created us in a way that makes us well suited for relationship with God and our neighbour. Whatever other shortcomings one may still find, enhancements will not provide any benefit in this regard. When looked at in this way, even a genetically 'improved' memory would not alter our relationship with God or make life any better in this context.[16] Creation 'in the image of God' then means that any such 'improvements' are no longer necessary. On the contrary, by focusing on them we might miss the fundamental dignity that is already inherent in the human condition. Judged by theological standards, the suggestion that enhancements are unnecessary should not be surprising. The New Testament notoriously disrupts our common values: God chose to reveal himself in the misery of a victim of torture (1 Cor. 1.23), and in tending to the vulnerable, we encounter Christ in them (Matt. 25.31–46). There may be an unexpected strength in shared vulnerability that protests against our tendency to exclude those at the margins.

The suggestion that, by using enhancements to reduce our vulnerability, we may lose something important about being human requires further consideration. Those who do not share Christian beliefs may not resonate with the theological claim that our created, frail nature is the ideal condition from which to be in relation with God. They might rather suggest that life in general would be better with genetically enhanced capacities. At this point the theological assertion about vulnerability raises an interesting question: are the promises of a more meaningful life through genetic enhancements realistic? Are there clues in other disciplines, ranging from literature to economic theory and psychology, as to whether 'more' really is 'better'? How do people with disabilities think about their life quality, for example?

These are profound questions! We should not hide from such difficult topics, for genome editing is a transformative technology that may change medicine – and perhaps even the way we think about our bodies. Among the many potential uses, there are helpful and relatively uncontroversial medical applications in children and adults. However, some people would also like legislators to pave the way for the genetic modification of human embryos, in order to alter their predisposition to disease, leading to the anticipated birth of a healthy child. Perhaps the birth of genetically modified embryos in China may be a sign of some of the things to come. Finally, as we have seen, some even hope that genome editing will 'improve' our lives by giving healthy bodies new and supposedly better capabilities, and well-respected working groups have argued that society should make forays in this direction.

This book aims to start a conversation on the ethics of genome editing. What uses should we allow or prohibit? What scenarios are realistic and what science do we need to know to enable us to have an intelligent discussion? How do we think of our bodies, and how do disease and physical capabilities figure in our view of life? In thinking about this far-reaching topic, it is crucial to take a close look at the arguments of both secular and religious people, scientists and thinkers of various persuasions to see differing views on what makes for a good life, well lived. What should we make of certain religious arguments, and can people of faith and of none agree on a common position? This is an exciting topic, and the debate requires looking at life from various angles.

Notes

1 Margaret Atwood, *Oryx and Crake* (London: Bloomsbury, 2003); Peter James, *Perfect People* (London: Macmillan, 2013).

2 Jennifer Doudna in the BBC World Service's HARDtalk with Sarah Montague, 1 June 2018.

3 Karl Barth, *The Church Dogmatics, Vol. III.4: The Command of God the Creator* (Edinburgh: T. & T. Clarke, 1961), pp. 363–4.

4 The Royal Society, 'UK public cautiously optimistic about genetic technologies', 7 March 2018, https://royalsociety.org/news/2018/03/genetic-technologies/ (accessed 25.9.2020).

5 National Academy of Sciences, National Academy of Medicine, *Human Genome Editing: Science, Ethics, and Governance* (2017), www.nap.edu/catalog/24623/human-genome-editing-science-ethics-and-governance (accessed 25.9.2020); Nuffield Council on Bioethics, *Genome Editing and Human Reproduction: Social and Ethical Issues* (2018), https://tinyurl.com/yy6pjgne (accessed 5.10.2020).

6 Nigel Biggar, *Behaving in Public: How to do Christian Ethics* (Grand Rapids, MI: Eerdmans, 2011), p. 59.

7 Information on sickle-cell disease and other genetic conditions can be found on the websites of the American National Institutes of Health, www.nhlbi.nih.gov, the UK's NHS or on the website Online Mendelian Inheritance in Man (OMIM), www.omim.org/.

8 Raúl Aguayo-Krauthausen, *Dachdecker wollte ich eh nicht werden: Das Leben aus der Rollstuhlperspektive*, 3rd edn (Reinbek: Rowohlt, 2015) (trans. A. Massmann).

9 See John Harris, 'Should we attempt to eradicate disability?' *Public Understanding of Science* 4:3 (1995), 233–42.

10 Francis Fukuyama, *Our Posthuman Future: Consequences of the Biotechnology Revolution* (New York: Picador, 2002), p. 10.

11 Tobias Hürter, 'Philosoph Julian Savulescu: "Ein Nasenspray kann die Beziehung festigen"', *Die Zeit*, 9 Sept. 2013, https://tinyurl.com/y6nplhrh (trans. A. Massmann).

12 Irenaeus actually wrote, 'God's glory is the living person', in *Against heresies*, IV.20.

13 Ted Peters, 'Can We Enhance the Imago Dei?', in *Human Identity at the Intersection of Science, Technology and Religion*, Nancey C. Murphy and Christopher C. Knight, eds (Farnham: Ashgate, 2010),

pp. 215–38; Ronald Cole-Turner, 'Religion, Genetics, and the Future', in *Design and Destiny: Jewish and Christian Perspectives on Human Germline Modification*, R. Cole-Turner, ed. (Cambridge, MA and London: MIT, 2008), pp. 201–23.

14 Harris, 'Should we attempt to eradicate disability?'.

15 Richard Dawkins, 'Afterword', in *What is Your Dangerous Idea?* John Brockman, ed. (London: Harper, 2007), pp. 297–301.

16 Gerald McKenny, *Biotechnology, Human Nature, and Christian Ethics* (Cambridge: Cambridge University Press, 2018), p. 147.

2

How does genome editing work?

Genome editing is a new technology that allows us to make very specific changes to a person's DNA, with an ease that was unheard of just fifteen or twenty years ago. Genome editing, otherwise known as 'gene editing', is likely to be one of the most important biotechnological innovations of the twenty-first century. It gives us opportunities to alleviate or cure some diseases that were previously impossible to treat. We might even be able to use it to modify or enhance some traits, in the absence of disease. Some people are enthusiastic about this, while others are worried about the new possibilities. Before we address the ethical questions that are raised by the technology, this chapter explains how genome editing works and explores what it may or may not be able to do. We'll begin by describing what genes are and what they do.

The importance of DNA

DNA is the molecule that contains our genetic information, which we inherit from our parents. Its elegant structure of a double helix, resembling a spiral staircase, was discovered by James Watson and Francis Crick in 1953. Soon after that groundbreaking discovery, scientists realized that some inherited medical conditions can be traced to changes in the DNA sequence, or the particular DNA composition, of an affected individual. The long DNA molecules are subdivided into many individual sections called genes, each of which has a specific function. Changes to DNA, which are called mutations, can change the function of an entire gene or completely disable it, often resulting in an illness. Sickle-cell anaemia was one of the first to be analysed, in which the exchange of a single DNA building block for another produces a change in the protein haemoglobin, causing a debilitating disease. Another example is haemophilia, in which the blood fails to clot properly. The 'madness of King George' may have been caused by a defect in the way the body processes a substance called porphyrin – hence the name of the condition, porphyria.[1] We now know of over 10,000 genetic conditions that arise from simple changes in DNA sequence. They often occur at single positions within DNA, but can also arise from expansions or rearrangements, in which a region may be swapped from one place to another, producing cancers or other complex diseases. Until recently, the only treatments available for many of these conditions were a controlled diet, changes in lifestyle or lifelong medication. Any thoughts of correcting the underlying errors in the DNA were the stuff of science fiction.

This all changed recently with the development of new series of highly accurate 'genome-editing' technologies. Around the year 2000, scientists described the first specialized genome-editing procedures, so-called ZFNs and TALENs,[2] and more recently these technologies have been taken to another level with the discovery of CRISPR-Cas9. As a result it is now possible to edit any DNA sequence very precisely, making specific changes to a person's genome with relatively little effort. This raises many questions about whether we should do this or for what purposes. In a human body, what is normal and what is diseased or unhealthy, and which genes, if any, might we consider modifying and at what stage of the person's development? But before we dive into these possibilities, we need to understand better the basic science that underpins the technology. What is DNA? What is its function? What effect does it have on my health or character? Those who are already familiar with the basics of genetic illness may like to skip ahead to the section on 'Complex gene interactions', to a later part of this chapter, or move on to the following chapter on the ethics of medical genome editing.

Zooming in on DNA

DNA is of great importance for the organization and the functioning of our cells and bodies. From a chemical perspective it is a large repeating polymer consisting of only four different chemical units, to which we give the letters A, G, C and T. This simple alphabet is used as a code for building a cell's components and the factors that control when these are made and used. Four letters may seem like a very simple and restricted alphabet, but there is an immense number of ways in which the letters can be arranged. Although there are only four of these letters ('bases'), there are 16 different combinations of two letters (AA, AT, AC, AG, GA, GG, GC, GT, TA, TT, TG, TC, CA, CT, CG and CC) and 64 combinations of three letters (AAA, AAG, AAC etc.). As the length of the sequence increases, the number of possible combinations becomes massive. The cells of our bodies each contain three billion letters, so the numbers are unbelievably large. The entirety of these letters in a person's cells is called the genome, and it is their precise sequence that defines our individual genomes. Each person has their own unique sequence (unless they are an identical twin), which differs from anyone else's by about one letter in every thousand (i.e. about three million differences in total). This sequence is faithfully copied with great precision every time each cell divides, with an extremely low error rate. Although there are many different types of cells in a person's body (different cell types in the liver, kidney, heart, brain etc.), they all possess the same DNA sequence, though different regions of the DNA are active in different cells and tissues.

We don't need to go into the details, but the DNA sequence acts as code both for producing proteins, which are the body's main structural components, and the enzymes that are responsible for our metabolism. The DNA also contains many control mechanisms for deciding when and where each of these is made. A gene is a region of DNA, which can be between a thousand and one million letters long and which codes for a particular protein.

Proteins are composed of long strings of chemical building

blocks called amino acids. There are 20 different amino acids, which are arranged in a very precise order for each protein. Proteins can be as short as only a few tens of amino acids long, or may be many thousands, depending on the particular protein. The largest known is titin, which contains 38,138 amino acids. The order and sequence of these amino acids is very important and determines the structure and activity of the protein, and single changes can have devastating effects. It is the DNA sequence that acts as a code for determining the sequence of amino acids, with three successive DNA letters generating a 'word' that specifies each amino acid. So single letter changes in the DNA can lead to changes in these 'words', thereby altering the protein's sequence, structure and function, with consequences for health and disease. However, it turns out that only about two percent of our DNA codes for proteins. The rest (which was once erroneously called 'junk') is involved in a range of control mechanisms, determining which genes are active (switched on) in which cells. Changes in these control mechanisms can compromise function, which gives rise to cancers or uncontrolled cell growth and division.

The complete human genome consists of three billion genetic letters, and this DNA is packaged into segments called chromosomes. We have 23 pairs of these chromosomes, each pair containing one chromosome from one's mother and one from one's father. Many genes are contained on each chromosome, and the stretches of DNA that code for proteins are often separated by other stretches that are involved in control or packaging (or with currently unknown functions). Our genomes contain information for about 21,000 proteins. Different parts are active in different cells, so the cells in your brain are not the same kind as those in your eye, gut or big toe. Although every cell in your body contains exactly the same DNA sequence, different parts of that sequence are active in different cells.

The pairs of chromosomes 1 to 22 are known as autosomal chromosomes, and each gene therefore appears twice, once on each of the paired chromosomes. This means that if one copy is damaged, then the other one can often compensate for the loss, and the individual may be free from disease, even though they are a carrier of the mutation. This type of genetic change is known as a 'recessive' condition and will only be manifest as a disease if both copies are incorrect. This is the case with several conditions; cystic fibrosis and sickle-cell anaemia are just two examples. A carrier may be perfectly healthy with one abnormal and one normal copy of the gene. That's one of the benefits of having two copies. There may be a problem, however, when two carriers have children. A child could of course inherit two normal versions of the gene in question, or one healthy and one damaged copy, and in both these cases the offspring will be free of disease. However, there is a one in four chance that the child will inherit two copies of the diseased gene, one from each parent, and will therefore have the condition.

Other genetic conditions appear even when only one copy of the relevant gene is damaged. These are called 'dominant' mutations, as the abnormal gene product dominates over the regular version and produces the symptoms. These are rarer than recessive mutations. The best-known example is Huntington's disease.

The twenty-third pair of chromosomes are the sex chromosomes, called X and Y. Females have two copies of the X chromosome, while males have one X and one Y. This means that if a gene is located on the X chromosome, then men only have one copy. If this is damaged then men have no backup, while women have a second one to 'mask' the mutated one. Examples of these types of genes are ones that cause some forms of colour blindness (which is more common in men) or those that lead to haemophilia and Duchenne muscular dystrophy.

Examples of damaged genes causing disease

Many thousands of DNA mutations have now been identified that lead to inherited, genetic diseases. In these over 10,000 so-called monogenic diseases, small changes in specific genes have profound consequences for the health and well-being of the individual. The best-known examples are thalassaemia, sickle-cell disease, haemophilia, phenylketonuria (PKU), Tay-

Sachs, porphyria, cystic fibrosis and Huntington's disease. Some of these arise from single letter changes in the DNA sequence, leading to single amino acid changes in a protein that render it inactive or with reduced function. One of the first to be described was sickle-cell disease, in which there is one letter change in the DNA sequence that codes for part of haemoglobin, the protein that carries oxygen in the blood. Sickle-cell disease is one of the most common inherited blood disorders, affecting hundreds of thousands of people worldwide.

Sickle-cell disease occurs when a person has two abnormal copies of the β-globin gene, which codes for one half of the protein haemoglobin, the oxygen carrier in our red blood cells. This gene is located on chromosome 11. The gene defect is a single letter change in the DNA (changing an A to a T in one position – which results in glutamic acid replacing valine at one position in the chain of amino acids, one of the 574 that make the protein). This mutation has only a very small effect on the ability of the protein to carry oxygen under many conditions. However, under low oxygen concentrations, during temperature changes or dehydration, the mutant protein changes in significant ways, triggering a sickle-cell crisis. As the protein polymerizes and precipitates, some blood vessels are blocked, and certain body functions are reduced. In these situations, sickle-cell patients feel significant pain for several days. These affected individuals have two compromised copies of the gene. Anyone with only one abnormal copy is a 'carrier' of the gene who does not usually have symptoms. When infected with malaria, however, carriers have less severe malaria symptoms than people with two or with no abnormal copies of the gene. That may explain why the condition is more common in people with recent ancestors who lived in malaria-infected areas.

In principle this genetic disease could be cured by simply replacing the T in the mutant gene with an A, as in the normal version of the gene, and indeed this has been tested with the CRISPR-Cas9 genome-editing system in cell lines and mice.[3] If this could be applied to early human embryos, then the mutation would be removed completely for future generations. However, the success rate for this type of correction is very low, as it is easier to inactivate a gene than to repair it. An alternative strategy for treating the condition in adults is to use genome editing to reactivate a form of haemoglobin that is present in the fetus, but which is usually switched off shortly after birth. With this method, several people were treated in clinical trials in 2019, and while the full results are still awaited, it seems the treatment works. One woman at least appears to have been cured.[4]

As well as these simple so-called point mutations, there are several other ways in which a DNA sequence can be altered to produce a genetic disease. In the most common form of cystic fibrosis, three consecutive DNA letters have been deleted, leading to the removal of one amino acid from the chain of the CFTR protein (cystic fibrosis transmembrane conductance regulator), which controls the way in which ions move across cell membranes. Just as in sickle-cell disease, the condition cystic fibrosis is only evident in a person who inherits two copies of the mutant CFTR gene, one copy from each parent; those with one mutant and one normal copy are healthy carriers. In the UK about 1 person in 25 is a carrier of the mutant gene, and the condition affects about 1 in every 2,500 births. An editing technology that could repair the region of DNA that is affected would provide a cure for cystic fibrosis. This has been demonstrated in cells in the laboratory.

Other forms of genetic disease arise from rearrangements or amplifications of larger regions of DNA. Huntington's disease is a neurodegenerative disease of this kind. The key symptoms are uncontrolled movements, emotional problems and loss of thinking ability. In contrast to many genetic diseases, the symptoms only begin to appear at some point after the 35th birthday, mildly at first but becoming progressively severe and resulting in premature death. This incurable condition is caused by a particular kind of mutation, an irregular 'expansion' of the letters CAG in a region of the huntingtin gene, which leads to the production of a toxic protein that contains long repeats of the amino acid glutamine. Healthy people have between 10 and 35 repeats of this CAG region, while sufferers have between 36 and 120 copies. It is an autosomal dominant disease, which means that a single defective gene copy will cause disease, regardless of the other copy. Everyone with this mutation will eventually suffer from the condition. Autosomal dominant conditions are rarer than recessive conditions, and Huntington's is the most common condition of this kind, with one affected person in about 10,000–20,000. Another example of an autosomal dominant disease caused by a repeat expansion is ALS. A repeat of the sequence GGGGCC in the C9orf72 gene causes ALS (a form of motor neuron disease) or a type of inherited dementia, with an age of onset typically between 30 and 70 years. Similarly, Friedreich's ataxia is caused by expansion of a GAA repeat in a region of the frataxin gene (located on chromosome 9). This is also an autosomal recessive disease, with a mean age of onset of 15 years that causes walking difficulty, a loss of sensation in the arms and legs and impaired speech that worsens over time. The number of GAA repeats in unaffected individuals ranges from 6 to 27 repeats, while patients have GAA repeat expansions ranging from 44 to 1,700 repeats.

For these diseases one could imagine that a method for cutting out the repeated region of the defective gene would alleviate the diseases, and indeed this is being tested. While these genetic mutations are deterministic (i.e. those with the incorrect DNA sequence will eventually suffer from the disease), the effects are late in onset, with several years of healthy life before the debilitating symptoms appear. DNA editing of human embryos with these mutations would therefore prevent the appearance of disease later in life, in contrast to conditions like sickle cell and cystic fibrosis, which are manifest from birth.

Other DNA changes produce effects that are less deterministic, but increase the likelihood of developing a disease. Many cancers fall into this type of DNA change. The best-known of these are arguably BRCA1 and BRCA2, the genes that, if mutated, increase the risk of developing breast cancer. The BRCA genes do not actually cause breast cancer; they are tumour suppressor genes that normally prevent cancer by helping to repair damaged DNA. About one in 400 people have a mutated form of the BRCA1 or BRCA2 genes that are less effective at repairing broken DNA, so people with these gene mutations are more likely to develop breast cancer at a younger age. About 60% of women with the BRCA1 mutation and 45% with the BRCA2 mutation will develop breast cancer before the age of 70.[5] The consequences of these gene mutations are therefore not deterministic but alter the probability of being diagnosed with cancer.

Complex gene interactions

Many conditions are caused by combinations of genes that work together, and the simple idea that some gene 'is for X' is not correct or appropriate. For example, more than 36 genes are known to contribute to type 2 diabetes;[6] height is affected by 697 small variations at 400 locations,[7] and differences in intelligence result from hundreds of genes.[8] Traits influenced by several or many genes rather than one are called polygenic. Yet the popular media sometimes confuses matters with simplified stories about 'genes for...', whether it's gluttony, infidelity, smoking, violence or politics. Of course, these are gross oversimplifications. There is no single 'gene for' any of these characteristics, though a range of gene combinations may exert an influence. We should also remember that genes alone do not determine many of our characteristics, and we are profoundly influenced by non-genetic, biographical factors, such as diet, education, exercise and stress.

In these instances a more recent development is to combine the results of studies with thousands of people looking at the differences in many genes in a complex array called a whole genome analysis (WGA). This produces something called a polygenic score, which has been used to predict probabilities of different traits from intelligence to psychiatric disorders.[9] While this is a successful predictor across large groups of people, it says much less about a particular individual. Genetic associations that hold for groups are not predictive of individual characteristics or behaviours. Many people carry the same genes, but they differ in their traits. 'To identify a genetic variant is more straightforward – but arguably less informative – than to characterize the complex environment of the individual', an editorial in *Nature* notes.[10] With regard to intelligence, the authors of a large genomic study concluded that even if all the genetic contributors to educational attainment were known, their effect would probably be overshadowed by other factors, such as the socio-economic and educational status of a child's family. In the words of a researcher in cognitive psychology: 'It would be irresponsible to look at a polygenic score and use it to make a prediction for a single individual.'[11] So although there are fictional prospects of using genome-editing technologies to produce or select human embryos with enhanced capabilities in complex traits such as intelligence, these would involve making simultaneous changes to very large numbers of genes. That seems like a genuine difficulty, even though CRISPR-Cas9 can be used to change several genes at the same time. The record to date is probably over sixty simultaneous changes in pig embryos. Nevertheless, many non-medical enhancements have been proposed, from an increase in height to greater muscle mass, a more powerful memory or increased intelligence, partly based on animal experiments, partly on the comparison of human genomes. While such interventions do not necessarily seem realistic today, it is of course possible that the science will catch up, so we will discuss enhancements in Chapter 4.

Does DNA control my destiny?

For some diseases, DNA mutations will have a profound effect on health and well-being. However, despite its fundamental importance for life, DNA is not a blueprint. It contains the information needed to produce proteins and to regulate their production, but genes are parts of complex systems. They do very little by themselves, and traits emerge from the interactions between genes and developmental and environmental factors. For example, think about the growth and development of a human embryo. The initial cell's DNA contains the basic information that, in combination with other molecules (in the egg cell) and the environment (the mother's uterus), facilitates the growth of the single cell (the combined sperm and egg) into a multitrillion-cell person. There is not a linear deterministic relationship between genes and physiology, and other factors play important roles that affect when and where each gene is active. We should therefore be careful to avoid a genetic reductionism or determinism, which sees genes as pulling all the strings in our lives, resulting in a form of genetic fatalism. Genes are very important, and both technical and moral questions surrounding genome editing need to be discussed carefully. Yet whatever one's conclusions, we must be cautious to avoid undue alarmism. Our genes may limit or add to our abilities, but we are much more than the sum of our genes, and it takes much more than genes to make a human person.

Genome editing: the mechanism

Which brings us to the subject of genome editing. Wouldn't it be good if we could simply cut out any defective genes and replace them? We have become accustomed to organ transplants; here is the molecular equivalent – DNA transplants. Until recently this was only a dream, but developments in molecular biology now allow this to be done. Techniques with the names ZFNs, TALENs or meganucleases were the first to be described around the year 2000, though these have now been superseded by the powerful technique CRISPR-Cas9. This was developed about 7 or 8 years ago, and it can be tuned to edit any gene at will. The procedure is relatively straightforward and cheap to develop, and so CRISPR-Cas9 has already become a widely used experimental tool in many research laboratories. Genome editing is regularly used in plants, bacteria, non-human animals or human cell lines.

The precise mechanism by which CRISPR-Cas9 works involves some elegant biochemistry and molecular biology, though the details need not concern us here. For those who want to know more, many accessible descriptions of CRISPR-Cas9 are available online, like Jennifer Doudna's 2015 TED-talk, 'How

CRISPR lets us edit our DNA'.[12] In basic outline, if you know the precise region of a gene that you want to modify, then the relevant portion of its sequence is incorporated within a piece of genetic information (RNA) that serves as a 'guide' for the genome-editing apparatus. Cas9 is a special protein that takes this guide and uses it to seek out exactly the same sequence within the entire genome. It then cuts the DNA at precisely that position. This in itself is a remarkable feat, like hunting for a needle in a haystack, as this sequence may only be about twenty letters long and this has to be found against the background of the other three billion letters in the entire genome. This break is then recognized by the cell's natural mechanisms for repair. Cells must respond to cuts in their DNA, as chromosome breaks are normally damaging. Such breaks can occur naturally, and cells contain several systems for correcting the damage. Depending on the intended modification, scientists exploit one of two such systems. In the simplest of these (called non-homologous end-joining), special enzymes simply join any ends together in a sort of fail-safe mechanism. This repair process is not accurate, and it produces very short deletions or insertions at the break point. As a consequence, the gene that was cut is inactivated in that particular cell. Genome editing is therefore fairly straightforward for inactivating an aberrant gene. However, cells also use another process to repair chromosomal breaks, called homologous recombination, in which the cellular machinery looks for another copy of the sequence and matches the break against this. In normal circumstances this second copy comes from the other paired chromosome and the damage is repaired without any mutation. However, it is possible to add another piece of DNA that spans across the gap, which is then used as a template for the repair process. In this way a mutated DNA sequence can be repaired or another gene can be inserted. This would lead to the correction of a genetic mutation or the insertion of extra material, resulting, for example, in a new or enhanced function. This process is less efficient, but it can be used to correct genetic mutations or to optimize or enhance the exact coding sequence of a gene.

Genome editing in children and adults or in embryos

Each human body contains about 30 trillion cells, each of which contains the same DNA sequence, which is slightly different for each individual. These are arranged into about 200 different cell types (with different cell types in the liver, kidney, brain, blood etc.), each of which uses a different combination of the many genes contained in their DNA, generating different structures and functions. These are called 'somatic' cells, which have arisen from a single fertilized egg cell, and through very many divisions have developed into a fully formed human (including the so-called germ cells – ovaries and testes or egg and sperm cells – that will be used to form the next generation). Performing gene modification on a whole body, whether a child or an adult, will therefore not be trivial, requiring changes in very many cells. This procedure is called somatic genome editing. Making changes to some of the cells in the target tissue may be sufficient to alleviate the symptoms of a disease, or even to cure it. A major difficulty here is that many organs, or a sufficient number of their cells, are inaccessible, and it is difficult to deliver CRISPR-Cas9 to the right destination. Nevertheless, this method still holds significant promise. Genetic diseases that affect the blood have been at the forefront of this research, and the eye is also accessible to this type of treatment, though the technology is being explored for treating a host of other medical conditions. As discussed in later chapters, this form of genome editing raises few ethical questions in principle, as it is limited to the particular consenting individual. However, this procedure can also be used to modify genes that are not related to any medical condition. If we say that genome editing of somatic cells, in children and adults, is morally permissible for medical reasons, then should we also allow it to be used for non-therapeutic enhancements? The egg and sperm cells, the germ cells, would not be modified, and no changes would be passed on to later generations. That can be seen as an advantage, but many would argue that the technology should only be used for treating medical conditions.

Genome modification of cells within whole bodies has obvious practical limitations. It would be much more effective to modify (edit) cells at an earlier stage in development (i.e. the early embryo), when it consists of only one or just a few cells. As this is the source from which all later cells derive, all the cells in the individual will then contain the repaired gene, including the germ cells that will pass on to the next generation. The condition will then be permanently repaired, not just for that individual but for all their offspring. This is so-called germline editing, which is currently illegal in almost all Western countries, including the USA, the UK and continental Europe. In the procedure, egg cells and sperm would be retrieved from the parents-to-be; the egg cells would be fertilized in a lab; the resulting embryos would be genetically modified, screened to check that the intended changes had been successful and then implanted in the woman's uterus, resulting in a pregnancy. In short, the procedure combines genetic modification of the embryo with in vitro fertilization (IVF).

Modifying harmful genetic mutations once and for all through germline editing might seem an ideal outcome, but it raises a multitude of ethical questions. What are the risks? Are there problems with performing a modification when the individual cannot give or withhold consent? Would the modification of an embryo imply a negative judgement about those whose genes have not been 'corrected' for the mutation? What are the standards by which decisions would be made – would we be imposing an artificial idea of what is 'normal'? If we draw a line between legitimate medical interventions and doubtful, non-medical 'improvements' or enhancements, how exactly should we spell out the distinction? These and many other issues will be discussed in later chapters.

Many people will be especially worried about the risks of germline editing. The molecular mechanisms are exquisitely precise, but there is always the possibility of accidentally changing DNA in regions other than the intended target. As we've already said, finding the correct sequences is a remarkable feat, looking for a sequence of 20 or 30 DNA letters from within the total genome of three billion letters. What if some other similar sequences are unintentionally modified as well, in so-called off-target events? This could introduce mutations at genes in other places and so generate new diseases. Indeed, some laboratory studies have suggested that there might be numerous such events, though others are less pessimistic. The technology is being continually improved and there are powerful new variations of CRISPR-Cas9 that greatly reduce the risks.[13] In its simplest form the technology first induces breaks in the chromosome at the site of the modification, which are then resealed. DNA breaks can be highly dangerous if they are not repaired or if they are misrepaired, and are the cause of many cancers. What if the cell's repair enzymes joined the wrong two ends together?

Even if all these off-target events are minimized, we know that genes do not work in isolation but are part of complex biological networks that are finely tuned and balanced. Modifying one gene might disrupt this network, with unintended consequences. The risk seems relatively low when 'resetting' a single point mutation, like the one leading to sickle-cell disease, as described above. The procedure would be much riskier if several genes are modified at the same time. In that case, intended changes in one pathway might contribute to other traits in a different or overlapping pathway and have an unpredictable effect. We will probably never be able to exclude the possibility entirely that a well-intentioned genetic intervention will do more harm than good. However, that is the case with all forms of conventional medicines, none of which are entirely risk-free, and often come with long lists of potential side effects. The risks may never be entirely ruled out, but they need to be outweighed by the benefits, and genome-editing procedures must cross the same hurdles as other medications. After all, reproduction by traditional methods is not without risk, and every natural conception generates a few tens of new mutations in each generation. But is risk the only important ethical question?

Gene therapy

We should also mention an older technology known as gene therapy, which is related to genome editing, though clearly different, but can easily be confused with it. 'Gene therapy' looks similar to 'genome editing' at first, but has a longer history. It involves the *addition* of a functional gene into the cells of a patient, so that this gene takes over the work of a mutated, dysfunctional one. Without the means to cut DNA at a well-defined position, this treatment lacks the precision of genome editing and leaves the dysfunctional gene in place. In gene therapy, the new gene is often introduced by putting it into a 'harmless' virus, which delivers it into the cell, where it is inserted at a random location into the genome. This was used about fifteen years ago to treat an immune deficiency in children ('bubble boy disease'). The children were cured, but some went on to develop leukaemia, which occurred as a result of the imprecise and variable location at which the new gene was inserted, causing an inappropriate activation of a nearby gene that produces cancer. The technology has since been improved using different viral vectors that include 'insulator' genes alongside the new gene, which stop adjacent genes from turning cancerous.[14] This type of gene therapy can also be used for treating other disorders. While the gene therapy treatments of immune deficiency and leukaemia were exploratory, and were done on a 'compassionate use' basis, the first two regular gene therapy prescriptions were licensed in the USA only in 2017, almost 30 years after the first research trials of gene therapy in humans. By contrast, it seems likely that the first genome-editing treatments will be licensed much faster than that. Gene therapy remains unsatisfactory because, in contrast to genome editing, it lacks precision. For that reason, it is not suited for the modification of embryos. A modification at a later stage presents fewer ethical problems, but it leaves the defective gene in place, so it will still be passed to future generations.

Is embryo editing necessary?

Are there other ways of achieving the same 'therapeutic' goal of preventing transmission of genetic diseases? In many cases it can be argued that genome editing is not necessary and that conventional techniques can be used to screen embryos to prevent transmission of an affected gene. Pre-implantation genetic diagnosis (PGD) is already practised for IVF to test for embryos with certain known genetic risks. For example, if both parents are cystic fibrosis carriers (i.e. both have one irregular and one regular copy of the relevant gene), then on average 25% of the embryos will contain only correct copies and 50% will be disease-free carriers. However, 25% will possess two faulty copies and will go on to develop the condition. By simple genetic testing it is possible to determine which embryos are which and then only use the healthy ones for IVF, with no need for any genome editing. This procedure is legal in many countries. If, instead, embryos are gene-edited with CRISPR, a pre-implantation screening will still be essential to discover which embryos have been successfully modified. So editing could well be unnecessary, provided there are enough embryos that are free of the mutation. However, there might be a few rare scenarios in which there are no unaffected embryos, and PGD would not be of any help. These might include instances where one parent has two copies of a mutated gene for a dominant disease or both parents have two copies each for a recessive genetic disease. We will discuss such scenarios in Chapter 3, where we will also address the question of whether PGD can be justified even though a few embryos are likely to be discarded.

Notes

1 Timothy M. Cox et al., 'King George III and Porphyria: An Elemental Hypothesis and Investigation', *The Lancet* 366 (2005): 332–5.
2 John Parrington, *Redesigning Life: How Genome Editing Will Transform the World* (Oxford: Oxford University Press, 2016).
3 Megan D. Hoban et al. 'CRISPR/Cas9-Mediated Correction of

the Sickle Mutation in Human CD34+ cells', *Molecular Therapy* 24 (2016): 1561–9.

4 CRISPR Therapeutics, 'CRISPR Therapeutics and Vertex Announce New Clinical Data for Investigational Gene-Editing Therapy CTX001', 12 June 2020, https://crisprtx.gcs-web.com/news-releases/news-release-details/crispr-therapeutics-and-vertex-announce-new-clinical-data.

5 Karoline B. Kuchenbaecker et al., 'Risks of Breast, Ovarian, and Contralateral Breast Cancer for BRCA1 and BRCA2 Mutation Carriers', *JAMA* 317 (2017): 2402–16.

6 Christian Herder and Michael Roden, 'Genetics of Type 2 Diabetes: Pathophysiologic and Clinical Relevance', *European Journal of Clinical Investigation* 41 (2011): 679–92.

7 Andrew R. Wood et al., 'Defining the Role of Common Variation in the Genomic and Biological Architecture of Adult Human Height', *Nature Genetics* 46 (2014): 1173–86.

8 Robert Plomin and Sophie von Stumm, 'The New Genetics of Intelligence', *Nature Reviews Genetics* 19 (2018): 148–59.

9 Robert Plomin, *Blueprint: How DNA Makes Us Who We Are* (London: Penguin, 2019).

10 Editorial, 'No Easy Answers', *Nature* 493 (2013): 133.

11 Eryka Check Hayden, 'Gene Variants linked to Success at School Prove Divisive', *Nature* 533 (2016): 154–5.

12 Jennifer Doudna, 'How CRISPR lets us edit our DNA', TED-talk (2015), https://tinyurl.com/yxxhxndx (accessed 26.9.2020).

13 Heidi Ledford, 'Super-precise New CRISPR Tool could Tackle a Plethora of Genetic Diseases', *Nature* 574 (2019): 464–5.

14 Ewelina Mamcarz et al., 'Lentiviral Gene Therapy Combined with Low-Dose Busulfan in Infants with SCID-X1', *New England Journal of Medicine* 380 (2019): 1525–34.

3

Medical opportunities: Therapeutic genome editing

What motivates medical and genetic researchers, especially in their work with genome editing, is probably the hope to fight disease. There have been many media reports on new ways in which CRISPR-Cas9 promises to prevent, cure or at least alleviate disease. This contrasts with the way some commentators roundly reject genome editing, mentioning it in the same breath as nuclear warfare and environmental pollution.

Numerous medical research projects using CRISPR-Cas9 are already showing promise, although the technology has been available for only a short time. At the time of writing, there are positive preliminary reports from clinical trials in which CRISPR treatments have been given to over 20 patients suffering from sickle-cell anaemia and beta-thalassaemia. According to estimates, over 270,000 babies are born with sickle-cell anaemia or sickle-cell disease worldwide every year. These are painful conditions that tend to reduce life expectancy. Genome editing also shows great promise in other medical applications. In 2015, a London hospital used genome editing to cure one-year-old Layla Richards and, later, another young girl, both of whom were about to die of leukaemia. The technology used for these modifications was a forerunner of CRISPR-Cas9, though the use of CRISPR will probably further improve the procedure. Researchers are expanding this approach, hoping to use CRISPR-Cas9 to treat other kinds of cancer as well.[1] Using genome editing to combat cancer, researchers are building on the use of an older, pre-genome-editing technology that has already cured thousands of cancer patients.[2] This is gene therapy, which, as discussed in Chapter 2, has two clear disadvantages compared with genome editing: in gene therapy, there is no control over where a new gene is inserted into the genome; and the old, defective gene remains in place.

In surveys in the UK and the USA, as well as in various expert recommendations, there seems to be little doubt that we should take advantage of the medical benefits that CRISPR promises. However, the crucial question is exactly how we should use genome editing in medicine. The medical procedures just described are done on children or adults. Researchers are also exploring ways to prevent severe disease by editing the genome of early embryos – so-called germline editing. An embryo would be modified genetically and transferred into the mother's womb, resulting in a pregnancy. Although this is still mainly a hypothetical possibility, it would be a more powerful approach, with more dramatic and longer-lasting consequences. In this way, researchers would be able to 'correct' the signature mutations that are responsible for severe heritable diseases with a single, limited intervention. This procedure would remove the root cause not only of sickle-cell disease or haemophilia but also of more severe diseases like cystic fibrosis, muscular dystrophy or Huntington's disease. These diseases are the so-called 'monogenic' diseases, in which patients have inherited a distinct genetic irregularity from one or both parents. Although these monogenic diseases are relatively rare, their effects can be severe and life-limiting, and they are otherwise incurable. Each of these diseases could be permanently treated by using genome editing to modify the dysfunctional genes at the embryo stage.

However, we need to keep a sense of perspective: genome editing, even when done in embryos, is not a magic bullet against all disease. Under the headline, 'The Future Without Limit', *National Geographic* magazine wrote: 'In the future, George Church believes, almost everything will be better because of genetics. If you have a medical problem, your doctor will be able to customize a treatment based on your specific DNA pattern.'[3] It almost doesn't need saying that most medical conditions are not caused by genetics and arise from accidents, lifestyle or infection. Genome editing offers no help against those conditions that do not have a clear genetic cause: injuries sustained in accidents, disabilities due to complications at birth and infectious diseases like the flu are currently well beyond the reach of this method. Less misleading, but still problematic, is overly optimistic talk about specific conditions concerning which we presently know comparatively little. One survey commissioned by the Royal Society – the UK's academy of sciences – asked people in 2018 if they were in favour of germline editing if it provided a cure for arthritis. However, arthritis is not a single condition and it has different overlapping causes. Genome editing, whether conducted on embryos or adults, is unlikely to have much effect on these conditions for most people in the foreseeable future. This is an example of the numerous conditions about which we presently know too little to hold out particular hopes; Parkinson's and schizophrenia are presumably among them as well, though genetic factors may play a significant role.

Cancer has been dubbed 'the emperor of all maladies', and there are significant genetic aspects to the disease. However, effective cancer prevention by means of germline editing seems unrealistic in most cases. Cancer is not a simple monogenic disease, and it is not determined by a simple genetic mutation. Not only are the genetic patterns underlying cancer much more complicated, but cancer generally depends on environmental and lifestyle factors as well as other unknown influences. In most cases the presence of a so-called 'cancer gene' gives an increased likelihood of suffering from the condition, compared to those who have the regular version of the same gene. We will therefore need to ask whether it is appropriate to modify, in an embryo, a gene that increases the risk of contracting an illness during adulthood. In this light we will later consider genetic contributions to breast cancer, which have received a lot of attention in the media.

For certain forms of Alzheimer's disease, no significant genetic causes are known. In Familial Alzheimer's, however, genetic factors increase the likelihood of suffering from this disease, alongside other causal factors. But again, these genetic factors describe a risk or likelihood rather than determining the outcome. Technically, germline editing could be used to restore the 'defective' gene to its 'healthy' form. However, the benefit of a reduced medical risk much later in life comes at the potential cost of unforeseen accidents or side effects, possibly with an immediate effect on the developing embryo.

Some of the concerns come from teething problems with the technology and the risks of unintended consequences or accidents. However, we suggest that risk is not the only ethical issue, and this chapter will raise further questions about germline editing that go beyond risks and uncertainty. For example, if societies allow germline editing for medical reasons, what are the criteria for deciding what is a severe illness or a lesser inconvenience? Would people with minor, unusual characteristics find themselves labelled as *ill*, although they consider themselves merely *different*? People use the terms normalization and stigmatization to describe social pressures on those who are different and who do not conform to the norm. There is a genuine possibility that good intentions and a potentially beneficial technology may have illiberal consequences that lead to further discrimination.

Genome editing in embryos, resulting in a pregnancy, is presently illegal in most countries. Scientists have conducted research on human embryos, which they then discarded after two weeks, as required by their licence. At the moment there is consensus in the scientific community that we are not ready yet to change an embryo's genes reliably. However, in late 2018, a rogue scientist from China announced the birth of two babies that he had modified as embryos. In doing so he broke several Chinese regulations on genome editing, including poorly informed parental consent, a lack of a medical need and no ethics committee approval. This drew near-unanimous condemnation from scientists across the world. The Chinese authorities have since sentenced him to prison for three years and imposed a hefty financial fine. It appears that the two children are doing well, although they will need to be monitored closely for many years as they develop and grow, to ensure that they don't suffer any side effects. His plan was to make the children resistant to HIV, and the procedure was performed in order to protect against infection, rather than to prevent a heritable disease.

The Chinese researcher's aim was to introduce a genetic mutation into an embryo that occurs naturally in a small minority of people in the West. If people have this mutation in both copies of the gene in question (called CCR5), then a crucial detail in the three-dimensional structure of a protein on the surface of the cells is modified so that HIV – the virus causing AIDS – can no longer gain entry into its target cells. People living with the mutation do not appear to suffer any known detrimental effects, though one study suggested that this type of resistance mechanism comes with a trade-off, increasing susceptibility to West Nile fever and influenza. The mutation occurs naturally in only a small minority of people in the West, where West Nile virus is rare; though the mutation is even rarer in other parts of the world where this virus is more common.

Most researchers would probably not list the prevention of HIV infection via germline editing as their first priority in developing medical genome-editing procedures – after all, there are other ways of preventing HIV infection. Moreover, compared with this, there is ongoing work to develop a CRISPR treatment for people who already have AIDS.[4]

Three genetic conditions have especially attracted the attention of researchers; Huntington's disease, cystic fibrosis and muscular dystrophy. Germline modification can seem like an attractive option to prevent these conditions, as they are caused by errors in single genes. However, it is extremely important for all the risks and safety issues to be sorted out before this is applied in a clinical setting; any accident would not only be lamentable for the individual involved but would also create a backlash against the technology. So currently a lot of people are working on the question of how to bring the error rate in genome editing down to an acceptable level. Many observers suggest that the question seems not to be *if* but *when* they will be successful and the technology will be available for clinical therapeutic use.

As a consequence, many scientists and commentators assume that the legal situation will need to be adjusted to make use of the full medical capabilities that CRISPR-Cas9 offers. In the wider population, a survey found that between 60 and 70% of Americans support the genetic modification of embryos if it prevents a significant impairment or reduces the risk of a severe disease like cancer.[5] In the UK, it appears that between 70 and 83% of the population favours genome editing in embryos even for curing non-life-threatening diseases.[6] The American Academies of the Sciences and of Medicine argued for this procedure in 2017, and in 2018 the prominent British Nuffield Council on Bioethics largely followed suit. Several theological ethicists are also in favour of germline editing at least for some purposes.[7]

What difference does a gene make?

For genome editing to make a difference in fighting disease, it is self-evident that the condition must have a genetic cause that is well understood scientifically. If parents know that they carry a gene that leads to a severe disease, they may consider genome editing for their children, in order to prevent them from inheriting the mutated gene and developing the disease. For many genetic conditions, a child will only have the disease if both parents pass on the mutated gene, but with some conditions, inheriting a single gene from either the mother or the father results in symptoms. Conditions like sickle-cell anaemia, cystic fibrosis, Tay-Sachs disease and beta-thalassaemia are examples of the first type and occur once in every 2,000 to 100,000 births. Huntington's disease is an example of the second type and affects about one or two people in every 20,000 births.

In contrast to these monogenic diseases, for other conditions that have a genetic component it is harder to predict whether a person will have the disease or how intense the symptoms will be, based on the presence of a single genetic mutation. For example, with ageing populations across the West, there is a lot of attention to dementia. Alzheimer's is one of its forms, and among its different varieties, a particular kind is more likely in people who have a certain form of the APOE gene. There are three variants (called alleles) of the APOE gene: APOE2, E3 and E4. Everyone has two copies of the gene and the combination determines your APOE 'genotype'. The E2 variant is the rarest form, and carrying one copy reduces the risk of developing Alzheimer's. APOE3 is the most common form and doesn't seem to influence risk. However, the APOE4 form, which is present in about 10–15% of people, increases the risk for Alzheimer's and lowers the age of onset. Having two copies of APOE4 increases the risk of developing Alzheimer's by 12 times.[8] Commercial services that analyse the genomes of paying customers may inform patrons about their so-called 'APOE4 status'. Would people in the future perhaps choose to 'fix' the gene in their children, using germline editing? Again, we must emphasize that these are probabilities that affect the *risks* of suffering from the disease; they are not certainties. Any approach to genome editing in this case will need to balance the risk of the procedure against the lifetime risk of the condition itself, which is only manifest later in life.

Alzheimer's and APOE4 constitute a good example of the many difficult questions associated with seemingly simple genome profiles. Making predictions about future health from one genetic marker involves many unknowns. Although having two copies of APOE4 does seem to increase Alzheimer's risk significantly, not even everyone with two copies will end up with dementia, and many people with dementia do not have that gene. There are other genetic variants that are associated with Alzheimer's, including changes in a gene called presenilin1. Other genes, nutrition, habits and unknown factors may compensate or exaggerate the effect of a seemingly harmful mutation. Further, the so-called 'risk gene' APOE4 has also been associated with lower rates of vitamin D deficiency and even with higher intelligence, among children living in Brazilian slums and elderly people in the Bolivian Amazon.[9] Jennifer Doudna, the leading scientist often credited with the discovery of CRISPR-Cas9, suggests that society may not want to streamline the human genome too much, as seemingly pointless mutations may turn out to be particularly helpful when faced with newly emerging diseases.

Some genetic patterns can also raise the likelihood of particular kinds of cancer. In most cases cancer is not inherited but develops unpredictably over a person's lifetime as a result of accumulated mutations. So cancer prevention with genome editing would not be straightforward, even if germline editing were technically and legally possible. However, some forms of cancer run in families and some people may have a genetic pattern that increases their risk of developing particular cancers. In many instances, lifestyle adjustments may have a greater impact than genome editing. Genetics played a crucial role when the actress Angelina Jolie decided to have a prophylactic double mastectomy, together with the removal of her ovaries and fallopian tubes, as a precaution against cancer.[10] She had discovered that she has a faulty version of the so-called BRCA1 gene (BRCA for breast cancer). In the USA and the UK, direct-to-consumer genotyping services like 23andMe and others have offered personalized information about potential BRCA mutations to their customers for several years. What would the news of a BRCA mutation mean in practical terms?

By itself, Jolie's BRCA mutation would not necessarily put her at a very high risk of cancer. Yet she had already lost her mother, a grandmother, a great-grandmother and other relatives to ovarian cancer. The genetic predispositions seem to be stronger in her family than in others, with a high risk of both ovarian and breast cancer. Readers with a family history like that may wish to discuss taking a genetic test with their GP. A genetic 'correction' of BRCA genes in adults, following adverse test results, does not seem to be scientifically possible. However, 'repairing' a BRCA gene in an embryo *could* be. Would people consider that option for their children in a hypothetical genome-editing future?

Preventative germline editing against breast cancer may sound tempting to some, and this is what several scientists imagine for the future. In the UK, about 12.5% of women are likely to be diagnosed with breast cancer at some point in their lives.[11] This proportion increases to a range from 60 to 90% for women with the genetic defect of the kind that Jolie has. On the other hand, it is still possible that a woman with this defective gene may have no female relatives on either side of the family who have been diagnosed with breast or ovarian cancer. The cancer risk may not be dramatically higher in her case: not all BRCA mutations are equal, and there may be additional heritable or compensating factors. So the likelihood of cancer would be lower than she might expect based solely on Jolie's story. With an irregular BRCA gene, the likelihood of ovarian cancer also increases, up to roughly 50%, but the risk levels without the mutation vary, according to lifestyle and other genetic factors.[12] Nevertheless, BRCA mutations play a significant role in only about 5% of all breast cancer cases, and most women with breast cancer do not have the BRCA mutation. In contrast to the 5% of cases with the BRCA mutation, it is estimated that about 25% of all cases could be prevented through lifestyle choices such as avoiding alcohol, maintaining a healthy weight and being physically active.[13] In fact, Jolie rightly relayed the advice of her surgeon: 'A positive BRCA test does not mean a leap to surgery' – and much less would it imply a need to modify an embryo.

Even if the option to repair BRCA genes with genome editing in an embryo was available, it would probably not be in great demand. Not only would this require IVF (in vitro fertilization) instead of natural conception, but media attention would probably lead people to overestimate the significance of BRCA irregularities. A sensible alternative to the hypothetical germline modification of an irregular BRCA gene would be to instil a healthy lifestyle in children, combined with increased checks for cancer. After all, if cancer develops, the chances of recovery are better the earlier it is diagnosed.

What genome editing could do in the clinic

The cases in which genetics determines the likelihood of getting cancer, whatever one's lifestyle, are relatively few. Most cancers result from random genetic aberrations that accumulate over a lifetime, and which are impossible to predict. In the wider fight against cancer, genome editing can make an important contribution to treating those who have developed cancer. In one therapeutic strategy, some of the cells from the patient's own immune system are removed and modified before being infused back into the body, so that they can now attack the tumour. So far, this process, which is complex and expensive, has had to be tailored to the individual, in contrast to an off-the-shelf generic drug. However, there is a good chance that the relative ease of performing CRISPR-Cas9 genome modification will make a big difference.

The main interest in medical uses of genome editing, however, is less about cancer than about so-called monogenic conditions. There are over 10,000 known genetic diseases with a fairly clear-cut genetic profile that have a very high likelihood of causing symptoms. Monogenic conditions affect fewer people than infectious diseases or cancers, and sometimes they are not severe and can be accommodated. However, there are several that are debilitating or even fatal, and most of these are incurable. As mentioned before, several clinical trials have used CRISPR in adults in the hope of curing sickle-cell disease and beta-thalassaemia. Several trials in the USA appear to have been successful, but more tests will be required before the treatment will be widely available.

Sickle-cell disease is just one out of several severe hereditary disorders that affect the blood, and for these conditions, CRISPR-Cas9 is a particularly useful tool. There are good chances that it will also help patients suffering from haemophilia. Targeting genetic disorders that affect the blood is promising, as the cells are fairly accessible, compared with those diseases that affect deep tissues. In addition, in this case the modification of only a small proportion of all the cells is sufficient to overcome the symptoms.

In late 2017, there was another remarkable 'first' involving genome editing, when researchers modified genes within the body of a man with Hunter syndrome. This is a rare X-linked recessive genetic disorder, affecting about 1 in 100,000 male births. Severe cases lead to death during the teenage years. There are treatment options in other cases, but in 2017, medical scientists announced the first ever whole-body human genome-editing therapy that permanently altered the DNA in a patient with Hunter syndrome.

Before any of these therapies are used on humans in a clinical setting, they are first tested in animal models of the disease. These laboratory experiments are showing promising results against several diseases, including a certain type of diabetes (even though this is a complex condition with many different underlying genetic causes) and phenylketonuria (PKU), a heritable disease that does lasting damage to motor skills and cognitive capacities unless the affected person strictly keeps a specialized diet. The fact that these procedures work in mice does not mean that they will necessarily work in humans, but it is at least an important first step.

Turning from procedures in children and adults to the embryo, which conditions besides the three mentioned already – muscular dystrophy, cystic fibrosis and Huntington's – would researchers tackle with germline editing? To begin with, there are the more common, milder heritable conditions. Newborn babies are routinely checked for congenital hypothyroidism, and treatment is relatively simple and effective. Familial hypercholesterolemia, which is a heritable condition that causes elevated levels of cholesterol, is even more common, but again symptoms can already be well managed medically. The use of germline editing in these cases seems unnecessary. By contrast, Sanfilippo syndrome, Lesch-Nyhan and Tay-Sachs disease are examples of relatively rare but much more severe conditions. They can be detected in a fetus during pregnancy, which often leads parents to choose abortion. Some may argue that genetic germline editing might reduce the number of abortions after such a diagnosis. For Tay-Sachs disease, haemophilia and sickle-cell disease, a number of testing schemes have been developed to inform partners whether they carry the version of the offending gene. In this way they are aware of the potential risks and can make informed decisions on whether to have children. Each of these can also be detected, before implantation, in pre-implantation genetic diagnosis (PGD) in order to select only the healthy embryos for implantation in IVF. This testing procedure would also be an essential part of a genome-editing procedure, in order to detect which embryos had been successfully modified, so it can be argued that there is no pressing need for genome-editing technologies to prevent these conditions. However, most people will not have access to genome editing or PGD, so many children will be born with these genetic diseases. This raises the question of whether there could be genome-editing treatments for children or adults (rather than embryos) with these conditions, and one early clinical trial of gene therapy for Tay-Sachs has shown some early success.[14]

Huntington's is another disease that presents a challenge for genome-editing treatments. Inheriting even one faulty version of the HTT (huntingtin) gene from just one parent is sufficient to cause the onset of illness at some point between the 30th and 50th birthday. Inheriting one copy of the healthy gene from either parent is not sufficient for staying healthy, as the mutated copy generates a toxic protein that accumulates in the brain. The condition affects around 5–10 individuals in every 100,000 people, and the main symptoms are reduced muscle control and memory lapses. Once the symptoms begin, they are progressive, cannot be alleviated, and result in premature death. Genetic tests have been available for several years, but given the lack of medical options, children with a parent suffering from Huntington's often agonize over whether they should take a test. Is the knowledge helpful or not? On the one hand, it removes any uncertainty and allows for realistic future planning. On the other hand, if the test shows that they carry the disease, with no hope of a cure, they have to live with that knowledge. One researcher, Alice Wexler, has contributed significantly to the development of the test. Aged 74, she recently revealed that she has the condition as well.[15] Many with the gene are also unsure whether to have children. Some who *may* have inherited the gene but don't know, report being uneasy about an experience as common as misplacing a key: is this perhaps the first symptom of the disease? When symptoms do set in, they may also include mood swings, anger and depression.

Genome editing would provide a clear opportunity to treat this disease. Correcting the mutant gene in an embryo would be a possibility, which would then stop its propagation into future generations. As already noted, this is not without considerable risk. Is there any hope of using this editing technology to inactivate the mutant gene in adults who suffer from the condition? There has been some evidence that this might work, in mouse models of the disease and in human cells grown in the laboratory.[16] However, there are obvious difficulties when transferring the process to a real person. The brain is a very challenging target for genetic therapies – with over 80 billion neuron cells, it has famously been called the most complex entity in the universe. An error in a medical intervention in the brain would also be devastating, as it is central to our bodily functioning and well-being.

First, there is the problem of delivery: how to get the enzymes for editing the genome into the brain. This is usually done by incorporating the modifying machinery within a harmless virus. Second, there is the issue of selectivity: to knock out the affected gene without damaging the healthy one. By some clever personalized design features, it seems that this specificity can be achieved. A third problem is to ensure that there are no undesired 'off-target' effects that alter other genes. This could indeed be a problem if the modifying machinery is left in the brain, increasing the probability that over time further unintended modifications might happen. In another clever twist, researchers used two modifying machines, one to inactivate the harmful huntingtin gene, and the other to target the gene-correcting machinery itself, thereby inactivating it after it has done its work.[17]

Besides genome editing, however, there are now promising new ways to halt the progression of Huntington's that are currently tested in the clinic. For the first time it appears possible to deliver a substance into the brain that breaks down the harmful protein causing dementia.[18] Potentially, a change in the genetic root cause is not the only way to tackle Huntington's medically.

The lure of genome editing: the fascination with Project Recode

A much more far-fetched and fanciful project for genome editing is to give people an extensive, system-wide genetic overhaul. This is the aim of Project Recode.[19] The project proposes that immunity to all viral infections across the board might be achieved by changing the very nature of our DNA genetic code. We would then use a different genetic dictionary from that used by viruses, which would therefore be unable to hijack our cellular machinery. This is almost certainly an unachievable dream, but dozens of researchers have added their names to the project team. The idea is more akin to science fiction as it would require hundreds of thousands of changes to our genomes. The details don't need to concern us here, but the very thought illustrates the powerful lure for researchers to think big and devise genome-editing projects of extremely ambitious proportions. Unlikely though this project may sound, today's science fiction could become tomorrow's science fact. There is no doubt that genome editing will be able to make highly valuable improvements to our health-care system in the future. It is precisely because of the importance that we naturally attach to our health that researchers might promise radical changes and improvement. However, in health care we may have to beware of ideologies that promise unhampered progress.

Medicine: a double-edged sword

The alleviation of medical conditions is an important goal from an ethical perspective. From a Jewish and a Christian viewpoint, humans are created in the image of God, and care for the sick and dying has always been significant in Christian practice. The image of God is a prominent concept in the early chapters of Genesis, the first book of the Bible. In part, this term implied a responsibility to act on God's behalf, to work against the forces of chaos in the world that are antagonistic to life. Judaism has traditionally promoted health care as well. This is also the spirit in which Jesus healed people. Such Judaeo-Christian traditions, to which of course secular efforts were added, have had a lasting impact on society in its wider commitment to physical integrity. In the Islamic world, health care has also been an area of lively discussion and practical innovations. Although the society we live in no longer understands itself as Christian, many of our culture's approaches to health care are deeply shaped by Christian principles. The atheist philosopher John Gray even argued that without the cultural impact of Christianity, based on scientific materialism alone, we would not affirm the dignity of every person, and aim to provide health care to everyone, whatever their station in life.[20]

New technologies raise profound questions about what is healthy, whether we appreciate people whose illnesses or impairments will not go away and how we treat those who might be different. We might sign the forms on a genome-editing treatment for a child without having sufficiently questioned whether an irregular condition should be changed. For example, George Church, a leading geneticist, has narcolepsy, a condition that makes him fall asleep involuntarily during the day. He uses techniques to avoid that, but he cannot sit and stay awake for long and he no longer drives a car. Yet Church sees the condition positively, suggesting that narcolepsy has boosted his creativity. 'I don't think [narcolepsy] is my only source of innovative ideas', he muses, 'but it doesn't hurt. I am surprised sometimes how quickly things come to me when I am in and out of sleep. I have solutions that might have otherwise taken days!' Genetic factors play into narcolepsy, though the condition cannot be reduced to them. Nevertheless, Church has wondered if he would get rid of the condition if he could. 'I probably wouldn't fix my narcolepsy, even if I could … It illustrates that we need to be very cautious about eliminating certain genes.'[21] While narcolepsy is typically considered an undesirable illness, Church thinks it is 'a feature, not a bug'.

What about conditions like achondroplasia and hypochondroplasia, which result in stunted growth due to particular genetic combinations? These do not necessarily cause pain or limit life in other ways. Often these conditions are considered a disease, but why should we not regard them as one variation within the diversity of human body shapes?

These examples take us right into a difficulty: there are undoubtedly conditions that many of us would want to avoid, but it is difficult to decide how to distinguish these from other milder things that we might not want to eliminate, even if we could. Where does regular variety and diversity end and where does debilitating disease begin? Could mild or even debilitating diseases come with other benefits that are often overlooked?

The point is not simply that health is good and illness is bad. Life is of course more than just physical functioning. Good health can often allow for a richer social life, given that our environment includes obstacles like steps and stairs, which pose a challenge to wheelchair users. Yet why don't we adjust our environment? This is not the only difficulty, however. Chronic illness excludes many people from wider participation in society. Everyday routines may take more time and energy for a less able person, physical limitations may exclude joining others in an activity and medication may be expensive. Here people with chronic illnesses or disabilities depend on society accommodating them. The availability of social support varies between countries and regions, but on the whole, society supports the inclusion of the medically ill up to a point.

However, society may sometimes exclude people precisely because they have a particular impairment. Havi Carel, for example, a secular philosopher, struggles with a severely restricted lung capacity, due to a rare condition called LAM. People leave her feeling isolated when they make little effort to conceal their horror or anxiety at her condition or when their pity is condescending. Her problem is not simply of a physical kind: 'The rules change when you are ill. You are, as Erving Goffman put it bluntly, stigmatized. You become an outsider to the world of the healthy, an offensive reminder of the ugly underbelly of life.'[22] The more we idealize health, the more we exclude those who are different and beyond a cure.

People with disabilities commonly identify with their condition, affirming and claiming it as part of who they are, and they confront exclusionary practices. The thinker Tom Shakespeare, himself with stunted growth due to achondroplasia, wouldn't want to have altered his own genes to be taller. 'People with disabilities are ... unlikely to be queuing up for genetic modification', he argues: 'their priority is to combat discrimination and prejudice. To "fix" a genetic variation that causes a rare disease may seem an obvious act of beneficence. But such intervention assumes that there is robust consensus about the boundaries between normal variation and disability. Contrary to the prevailing assumption, most people with disabilities report a quality of life that is equivalent to that of non-disabled people ... and the voices of people living with illness and impairment need to be heard.'[23]

He points to many studies showing that people with disabilities are just as able to attain life satisfaction as others. This empirical perspective contrasts with the view by the philosopher Jonathan Glover, who writes that 'disabilities should be understood as obstacles to flourishing'.[24] An impairment, Glover suggests, does not necessarily reduce life fulfilment, but makes it harder to achieve. That might seem intuitive at first, but in reality it is misleading, as usually people adapt very well to their situation in life. While some conditions, such as depression, do indeed make a fulfilled life more difficult, this is not true for every impairment. The idea from utilitarian philosophy that greater health or more resources will give a greater chance of happiness across the board is simply not true.

The idea that parents should edit out characteristics that are considered debilitating goes against this drive towards inclusion and could create a harsher social climate for everyone. Rosemarie Garland-Thomson, a scholar and disability activist, comments that no one can avoid disability. Illness, accidents and age-related decline will touch us all eventually: 'At our peril, we are right now trying to decide what ways of being in the world ought to be eliminated.'[25]

Genome editing may soon be able to alleviate symptoms of particular diseases and impairments. However, for many other conditions, medical improvement is not on the horizon. Optimists will of course say that the difficulty of the task should not deter us from the search for new medical approaches. Perhaps at some point there will indeed be medical progress for these conditions. However, the crucial point is that medical and scientific progress involves a trade-off: when medical diagnosis becomes more precise, encompasses more conditions and becomes more readily available, the social meaning of the condition also changes.

One example of this social dynamic comes from Denmark, where births of children with Down's syndrome have long been in decline. Some 97% of pregnant women use the free pre-natal tests that the state has sponsored since 2004, and 95% of babies diagnosed with Down's are aborted. Down's is due to an extra copy of an entire chromosome, number 21. Genome editing, by contrast, works on the much smaller scale of individual genes, so it is very unlikely that parents who are worried by a Down's diagnosis will consider a genome-editing procedure. Nevertheless, the connection between Down's and genome editing is that a society's push to reduce impairment and disease with genome editing can make parents even less prepared to embrace a child with Down's. One Danish mother looked back at her surprise seeing that her newborn baby had Down's, and remembers her difficulties accepting the child. Had she known before, she would have aborted: 'That's the message I've been inculcated with: that Down's syndrome is something utterly terrible.'[26] Similarly, in the UK, it's commonplace for newspapers to use the word *risk* for the possibility of having a child with Down's, with the implication that this is undesirable and bad. This contrasts with studies that confirm high rates of life satisfaction among people with Down's, their siblings and their parents.[27] Many people with the condition work professionally, enjoy a significant degree of independence and are highly valued members of their communities.

The kind of stigmatization and marginalization that Havi Carel laments is likely to continue to increase as the flip side of medical progress. In the meantime, people with stunted growth like Tom Shakespeare do not, or not necessarily, suffer physically from their condition. Imagine, on the other hand, how the parents of a child with stunted growth may worry about the reactions that their child will experience in the social environment. There is a trade-off between expanding medical efforts and social acceptance of people with illness or impairment. The popular use of war metaphors for medical efforts, in newspapers and documentaries, is often not helpful either – the war on disease, cancer, obesity and so on. The unintended side effect of this fight against disease is that it reduces the ways in which people make sense of their conditions when they have incurable diseases. The concept of a fight against disease is not helpful for them, as they have a condition that society tries to prevent, and they need to find different ways of coping with this. The warfare script makes it harder for them to find well-being *in* sickness or to explore ways in which some aspects of their disease might improve life. They require resilience in embracing their condition and overcoming obstacles, but that can be either encouraged or discouraged by society. By focusing too strongly on medical cures, society does not make that task easier. Precisely because medicine 'works', it also has a 'pathogenic' effect, making things worse for those to whom medicine gives little help.

A moving example was an article on Ruth Weiss, a girl with a form of albinism that sharply reduces her visual acuity, making her legally blind.[28] Despite this limitation, she plays basketball for her local team and practises with great determination. When her father asked her if she wished that her parents had corrected the gene that caused her blindness before she was born, she said *no*. She gave the same negative response when asked if she would consider modifying the genome of her future children. Her father, a physician-scientist working at an American University, commented that he and his wife would have welcomed the opportunity to edit the blindness gene, had they been presented with that possibility before Ruth was born. But he now thinks differently and considers that doing so could have changed some of the things that make her special, and that 'would have made us and her different in a way that we would have regretted'.

Instances like this force us to think carefully about the implications of medical genome editing, especially in embryos. At the very least, the voices of those who are affected need to be heard. Some will be eager to push forward with the technology to eliminate debilitating conditions, while others are much more cautious.

This does not mean that society should stop investing in medical progress. A strict objection to all medical development would give up too easily on the possibility that people with, say, cystic fibrosis could enjoy several extra years in better health. There are, however, some conditions that consistently seem to reduce life fulfilment. These are depression, strong persistent pain and incontinence.[29] Being constantly bedridden may also reduce life satisfaction for many people. So in these instances the medical uses of genome editing could be highly beneficial. However, an all-out commitment to 'the forward march of science' would make us shut our eyes to the 'collateral damage' of social exclusion. From a Christian perspective, those who are left behind by medical progress or who are excluded and looked down upon are among 'the least of these' (Matt. 25.31–46). The Gospels summon Christians to attend to them with care, not out of condescension but because they too are made in the image of God and Christians encounter Christ in them – unbeknown to everyone involved.

Rather than understanding medical progress as an all-out fight against disease, we may need to face up to the collateral damage of stigmatization that can result from medical progress. In our opinion, measured medical progress should go hand in hand with a renewed, visible public commitment to support people with chronic diseases or disabilities. We may not want to pull out all the stops in our medical efforts. The pursuit of medicine needs to walk a fine line here: improvements in medical technologies must not disadvantage or discriminate against those who are beyond the reach of medicine; and with certain conditions, even medical 'success' does not make life better. Society should take those with incurable conditions seriously and support them, both in practical and symbolic ways, and we need to be open to hear the voices of those for whom an impairment contributes to the quality of life. If not, then our legitimate concern for health could have harmful consequences.

For that reason, it is particularly dangerous to idealize genome editing as the ultimate weapon in the fight against disease. If we are willing to learn this difficult lesson, then improvements in health care with genome editing can be morally helpful and important.

Modifying embryos to prevent disease? PGD as an alternative

One question that results from this reflection is whether we should support medical germline editing. Here an important point is that pre-implantation genetic diagnosis (PGD) can usually be considered as an alternative to the genetic modification of embryos. A couple who are concerned about passing on a genetic disease to their child may choose in vitro fertilization (IVF) with a crucial additional step. In IVF, a lab fertilizes egg cells from the woman with the man's sperm in a petri dish, so that a few embryos result, one of which, or a few, would then be implanted in the woman's uterus. In contrast to genetic germline modification, there is no modification of the embryo's genes before implantation in PGD; rather, the embryos are screened so as to exclude those with a disease mutation, selecting to implant only the 'healthy' ones. For example, this procedure is used by parents who wish to avoid passing on a genetic predisposition to Huntington's disease or cystic fibrosis. In Cyprus, PGD has been used to test for beta-thalassaemia, a severe blood disorder that is unusually common on the island, with about one case in 1,000 live births.

We might ask if the medical benefits of PGD are any different from those of germline editing. Those who are concerned with the stigmatizing effects of our pursuit of medical progress will question whether PGD offers any advantage over germline editing. Both procedures are designed to exclude disease, rather than to learn to live with it. Moreover, some people criticize PGD as it results in the destruction of the unwanted human embryos. They suggest that the procedure infringes upon the human dignity of the discarded embryos. These arguments apply equally to PGD and genome editing, as they both require embryo selection. However, we argue that rather than permitting germline editing, we should regulate PGD in a way that encourages its good use.

PGD is legal in the UK, the USA and in most European countries. In Europe and the UK, regulations prevent the use of PGD to select for non-medical traits such as the embryo's sex. PGD has a lower technical risk than genome editing, and its social acceptance is higher. Changing an embryo's genome in order to select for non-medical characteristics will require much wider debate (see the next chapter). There are fewer options to do so with PGD, in which it is only possible to select for or against genes that are already present and have been inherited from one or other parent. Attempts to increase physical height or intelligence by selecting for a particular genetic profile will be ineffective, as these result from the effects of combination of many genes, and the 'desired' combination may not be present within the limited number of embryos. Since PGD is legal and morally defensible, there is a strong argument for restricting genome editing to its uses in children and adults, where it can offer highly significant benefits.

Nevertheless, even PGD results in discarding the unwanted human embryos. These will include the unselected ones containing the mutations, along with any surplus 'healthy' embryos, though these will be fewer in number than in IVF. As with IVF, some implanted embryos will fail to implant or to develop. If human life begins at the moment an egg is fertilized, as some people argue, then a living human embryo must be given all the protection that is due to a human person, regardless of whether it has been implanted in the mother's uterus.

However, we need to consider whether this argument is ultimately persuasive. It is estimated that at least a third of all egg cells that are fertilized after natural intercourse fail to implant in the uterus and are expelled from the woman's body.[30] If fertilization has taken place but the embryo failed to implant, then this natural process would have involved discarding a human person, if that is indeed what an early embryo is. This scenario differs from the rarer one of miscarriage, in which an embryo or fetus, which has already implanted in the womb and developed further, dies. Some may argue that when comparing embryos that fail to implant after natural intercourse with those lost in PGD, the lack of intention is the important distinguishing moral factor. Yet that cannot be the real point once we have looked at the numbers. For every two babies who have ever grown in their mother's womb, another one person would have died without anyone even noticing, due to the embryo's failure to implant – if indeed every embryo is a person in the full sense. If every human embryo has full human dignity immediately after fertilization, then nature presents us with a massive, unimagined loss of human life.

Note that the moral issue here is not abortion; though it is related to the question of when a fully human life begins. Here, with PGD, we are concerned with embryos that are no more than a few days old, which have not yet implanted in the mother's womb. The early embryo, consisting of a few cells with no structure or nervous system, is not yet recognizably human. There are therefore highly significant differences between the moral questions of PGD and abortion. One Christian ethicist suggests that the high number of embryos that are lost naturally before implantation is not a sufficient reason to allow IVF and PGD; instead, we are told, this should be a cause for lament and grief.[31] Critics of abortion do lament the fetuses that are killed in abortion, for example in special liturgies and public prayers. However, no corresponding statements are to be found for the *early* embryos that are lost naturally, after 'natural' intercourse, before implantation – even though their numbers are much higher. It appears that at the end of the day, to attribute full human dignity to an embryo from day one, whether it has implanted in the mother's uterus or not, is not persuasive – even to many strict traditionalists.

Of course, this does not mean that we should dispose of early human embryos as we would of any other material object. Embryos do require protection, especially in the face of the amazing way in which, in the right environment, an embryo of just a few cells can grow into a wondrously complex human body. Even if the protection of early embryos does not need to be absolute under all circumstances, a valid reason should be required before performing PGD.

When discussing the moral status of early embryos, it is necessary to take a more responsive approach rather than adopting an uncompromising position on their protection. It seems that the destruction of a small number of *early* embryos in PGD can be tolerated *if* it allows for the prevention of a severe disease in a child.

Does that then imply a blanket justification of PGD under any circumstance? No. PGD should be limited to medical considerations, though we still need to define what is a *significant* medical justification. We also need to ask whether the simpler process of PGD is an alternative to the medical genome modification of embryos. In most cases it will be. The issue is whether there are enough unedited embryos that are free from the harmful mutation to attempt implantation, which may require more than one attempt before it results in a pregnancy. This number will be smaller if it is necessary to exclude *several* harmful mutations simultaneously, or if both parents carry a defective dominant gene. These cases are rare, as germline editing advocates concede. Sometimes it may also be possible to select sperm without the harmful mutation before fertilization, resulting in a greater number of eligible embryos. In practice, this process will usually be for the purpose of excluding specific monogenic diseases, in which a clearly defined mutation causes a severe disease, and in these instances PGD will usually make germline genome editing unnecessary.

However, one might argue that in one sense the therapeutic modification of embryos is preferable to PGD. PGD is a 'negative' process as it deselects and excludes an embryo from a pregnancy, while genome editing 'repairs' and restores rather than discards – though any unedited or damaged embryos as well as potential unimplanted, modified embryos will still be excluded and discarded. However, PGD is technically much less risky and it does not open the door for the creation of non-medical characteristics.

A more important criticism of PGD is that, by selecting between embryos, we are making a judgement that life with a medical condition is not worth living or is less valuable than a healthy life. However, life with a chronic medical condition is not worth less than one without. We cannot even say that a medical condition will necessarily lead to a less fulfilling life. In quality-of-life research, several studies have shown that on the whole, people with disability are very satisfied with their life quality, though a meaningful and fulfilled life will be made more difficult by conditions that cause significant and persistent pain. The parental perspective and context will also make an important difference. The Christian author Ellen Painter Dollar explains why PGD was the right choice for her when having her second child. Her first child had inherited a disabling genetic bone disorder, osteogenesis imperfecta. The condition did not diminish her love of the first child, but when she wanted a second, she felt that taking care of another child with disability would be beyond her powers.[32]

In Europe, PGD is legal, with the restriction that it can only be used to prevent severe conditions. This contrasts with the situation in the USA, where the procedure can unfortunately even be used to select the embryo's sex. In the UK, its use is governed by the Human Fertilisation and Embryology Act (HFEA), under which PGD is not acceptable for 'social or psychological characteristics, normal physical variations, or any other conditions which are not associated with disability or a serious medical condition'. Sex selection is allowed only as a method to exclude a disorder that affects one sex but not the other.

Of course, PGD shares in the dilemma sketched out for all medical innovations: with the technological reduction of disease, the marginalization of those with disease is likely to increase. There does not appear to be an easy solution to this problem. For that reason, PGD is best restricted to excluding a very severe disease. Having the Huntington's gene would typically be recognized as a justification for PGD. Such a rule should, however, be accompanied by a renewed public commitment to include people with an impairment more effectively in public life. Genome editing *in children and adults* also serves the goal of physical well-being, which has a general acceptance in various religious and secular moral traditions. However, our moral recommendation in the medical area takes the form of a compromise. From a Christian perspective, we are called to take the concerns even of 'the least of these' very seriously. However, the examples of where medical progress is not necessarily the same as moral progress do not merely rest on religious convictions. For that reason, genome editing in children and adults should go together with a renewed effort to include people with chronic illness and disabilities in society. *Germline* editing, on the other hand, may soon overcome its technical difficulties and risks, but we generally reject it on other, moral and practical grounds.

Notes

1 Jennifer Couzin-Frankel, 'CRISPR takes on Cancer', *Science* 367, issue 6478 (2020): 616.

2 Charles Graeber, *The Breakthrough: Immunotherapy and the Race to Cure Cancer* (London: Scribe, 2018), p. 176.

3 Peter Miller, 'George Church: The Future without Limit', *National Geographic*, 2 June 2014, https://tinyurl.com/y6qn55dq (accessed 26.9.2020).

4 Rob Stein, 'CRISPR Gene Editing may offer Path to Cure for HIV, First Published Report Shows', *NPR*, 11 Sept. 2019, https://tinyurl.com/yyf7pzkf (accessed 24.9.2020).

5 AP-NORC poll Dec. 2018, https://apnorc.org/projects/human-gen etic-engineering (accessed 24.9.2020).

6 A survey asked in 2017 what respondents thought of 'heritable' genetic changes, not mentioning embryos: Anita van Mil, Henrietta Hopkins and Suzannah Kinsella, 'Potential Uses for Genetic Technol-ogies: Dialogue and Engagement Research conducted on Behalf of the Royal Society', https://tinyurl.com/y4q4dknv, see also https://tinyurl.com/yddl4pqh (accessed 24.9.2020).

7 Ronald Cole-Turner, 'Religion, Genetics, and the Future', in *Design and Destiny: Jewish and Christian Perspectives on Human Germ-line Modification* (Cambridge, MA and London: MIT Press, 2008), pp. 201–23.

8 Daniel M. Michaelson, 'APOE ε4: The most Prevalent yet Under-studied Risk Factor for Alzheimer's Disease', *Alzheimer's & Dementia* 10 (2014): 861–8.

9 Moises Velasquez-Manoff, 'The Upside of Bad Genes', *The New York Times*, 17 June 2017, https://tinyurl.com/y8ggcnlv (accessed 24.9.2020).

10 Angelina Jolie Pitt, 'Diary of a Surgery', *The New York Times*, 24 March 2015, https://tinyurl.com/y6gdocwy (accessed 26.9.2020).

11 Henry Scowcroft, 'Angelina Jolie, Inherited Breast Cancer and the BRCA1 Gene', Cancer Research UK Science blog, 14 May 2013, https://tinyurl.com/y5xsdcc2 (accessed 24.9.2020).

12 Memorial Sloan Kettering Cancer Center, 'BRCA1 & BRCA2 Genes: Risk for Breast and Ovarian Cancer', https://tinyurl.com/y3js weqe (accessed 24.9.2020).

13 For statistics like these, see the websites of UK Cancer Research (www.cancerresearchuk.org/), the American National Institutes of Health (www.nih.gov/), and the American Cancer Society (www.cancer.org/).

14 Carey Goldberg, 'Early Report: Baby Treated with Gene Therapy for Deadly Tay-Sachs Disease appears to Stabilize', WBUR, 24 Oct. 2019, https://tinyurl.com/y2usnacv (accessed 24.9.2020).

15 Denise Grady, 'Haunted by a Gene', *The New York Times*, 10 March 2020, https://tinyurl.com/y5c3klcv (accessed 24.9.2020).

16 Alex Mas Monteys et al., 'CRISPR/Cas9 Editing of the Mutant Huntingtin Allele in Vitro and in Vivo', *Molecular Therapy* 25 (2017): 12–23.

17 Nicolas Merienne et al., 'The Self-Inactivating KamiCas9 System for the Editing of CNS Disease Genes', *Cell Reports* 20 (2017): 2980–91.

18 Hannah Devlin, 'Excitement as Trial shows Huntington's Drug could slow Progress of Disease', *The Guardian*, 11 Dec. 2017, https://tinyurl.com/ybfcxkzr (accessed 24.9.2020).

19 Jef D. Boeke et al., 'The Genome Project-Write', *Science* 353, issue 6295 (2016): 126–27.

20 Michael Prodger, 'Matters of Life and Death: Rowan Williams and John Gray in Conversation', *New Statesman*, 30 Nov.–6 Dec. 2018, 46–51.

21 Zoë Corbyn, 'George Church: "Genome Sequencing is like the Internet back in the late 1980s"', *The Observer*, 18 Feb. 2018, theguardian.com/science/2018/feb/18/professor-george-church-nebula-genomics-interview (accessed 24.9.2020).

22 Havi Carel, 'My 10-year Death Sentence', *The Independent*, 19 March 2007, https://tinyurl.com/y6n5v8vw (accessed 24.9.2020).

23 Tom Shakespeare, 'Gene Editing: Heed Disability Views', *Nature* 527 (2015): 446.

24 Jonathan Glover, *Choosing Children: Genes, Disability, and Design* (Oxford: Oxford University Press, 2008), p. 88.

25 Erika Check Hayden, 'Should you edit your Children's Genes?', *Nature* 530 (2016), 402–5.

26 Maria Stöhr, 'Wo es Kinder wie Daniel bald nicht mehr geben könnte', *Spiegel Online*, 24 Dec. 2019, https://tinyurl.com/y5y398xe (accessed 24.9.2020) (trans. A. Massmann).

27 See the studies by Brian G. Skotko, Susan P. Levine and Richard D. Goldstein: 'Having a Son or Daughter with Down Syndrome: Perspectives from Mothers and Fathers'; 'Having a Brother or Sister with Down Syndrome: Perspectives from Siblings'; 'Self-perceptions from People with Down Syndrome', all in: *American Journal of Medical Genetics Part A* 155 (2011), issue 10; Skotko et al., 'Family Perspectives about Down Syndrome', *American Journal of Medical Genetics Part A* 170 (2016), 930–41. See also Gary L. Albrecht and Patrick J. Devlieger, 'The Disability Paradox: High Quality of Life against all Odds', *Social Science & Medicine* 48 (1998), 977–88.

28 Hayden, 'Should you edit your Children's Genes?',.

29 Daniel Kahneman, *Thinking, Fast and Slow* (London: Penguin, 2012).

30 N. S. Macklon, J. P. M. Geraedts and B. C. J. M. Fauser, 'Conception to Ongoing Pregnancy: The "Black Box" of Early Pregnancy Loss', *Human Reproduction Update* 8 (2002): 333–43.

31 Robert Song, *Human Genetics: Fabricating the Future* (Cleveland, OH: Pilgrim Press, 2002).

32 Ellen Painter Dollar, *No Easy Choice: A Story of Disability, Parenthood, and Faith in an Age of Advanced Reproduction* (Louisville, KY: Westminster John Knox Press, 2012).

4

Cornucopia or Pandora's box? Genetic enhancements

Introduction

Life is unfair! Born with the right genes for sports, some people win the genetic lottery, and with the right training regimes, they take home all the trophies. The Finnish skiing legend Eero Mäntyranta had an unusual genetic mutation that gave him greater endurance, to the chagrin of many who regularly finished behind him. Mäntyranta had a condition called primary familial and congenital polycythaemia (PFCP), which is an inherited mutation in the erythropoietin (EPO) receptor gene. The mutation increased the number of his red blood cells and with it the oxygen-carrying capacity of his blood increased by up to 50% – quite some advantage in endurance sports! There are many other examples of athletes who were born with the right genes for athletic prowess: marathon runners like Eliud Kipchoge, who comes from the Kalenjin tribe in Kenya and appears to have a favourable combination of genetic endowments. Caster Semenya, the successful South African middle-distance runner, has unusual levels of testosterone. Similar differences, though with multiple genetic components working together in a much more complicated way, may also apply to traits like intelligence, height, musical ability, character or any number of things. It seems unfair that some people are born with inherent advantages over others; should we maybe try to level the playing field?

It is an uncomfortable fact that we are born with different inherent abilities, though we like to think that excellence is something that we achieve by our own efforts, not by virtue of our inheritance. We often talk of people having a 'gift' or talent for something, yet this presents a problem for egalitarian societies for which success should be due to merit, not ancestry. In the opinion of one vocal group of thinkers, this is a serious moral cause that urgently needs to be righted. They've come up with their own suggestion for levelling the playing field: in the future, genome editing could give parents the option to 'upgrade' the athletic prowess of their children, potentially giving equal opportunity to everyone. They even contend that, if done safely, this legalized 'gene doping' would do away with clandestine doping, and sports would become safer and fairer. Moreover, if genes are involved in giving some an advantage over others by nature in various areas, not just in sport, genetic enhancements could seem to improve the lives of many people. Enhancements could seem like the cornucopia of mythology, providing us with an unexpected abundance of benefits.

Some types of genetic enhancements for adult athletes are already technically possible. For example, a gene therapy called Repoxygen was devised for people who suffer from anaemia. Their blood carries too little oxygen to sustain regular levels of physical activity. To correct for that, this therapy inserted a gene into cells in the person's body, to boost the production of the hormone EPO, which in turn increases the amount of oxygen that can be carried in the blood. This strategy was abandoned for treating anaemia in 2003. However, in sport, higher blood oxygen allows for greater endurance, with benefits for cyclists, long-distance runners and cross-country skiers. Another way to boost the oxygen-carrying capacity is by blood doping, with athletes taking transfusions of their own blood that has been taken and stored before an event, adding to their red-blood-cell count. Alternatively, injections of EPO itself will increase the number of red blood cells. Such blood doping is illegal in professional sports, where anti-doping agencies use tests to identify the culprits. In contrast, a gene therapy approach would remove the need for cumbersome blood transfusions or drug injections. To introduce Mäntyranta's EPO mutation into an embryo would be even more far-reaching.

This would endow the child with this extra potential from birth. Such an enhancement wouldn't be totally 'artificial', as it would simply redistribute genes already occurring within the natural gene pool. For that reason it would be difficult to detect whose genes were genetically modified. So some people argue that we should relax anti-doping rules, which currently are not particularly effective anyway. Instead, the suggestion goes, we should make gene doping widely available, so that everyone has the same advantages.

Some genetic enhancements seem far-fetched and are closer to the realm of science fiction. Genetic modifications to facilitate travel in outer space have been suggested, as too have methods for improving night vision or genetic modifications for military combat.[1] We know that our genomes contain many inactive genes (so-called pseudogenes) from our evolutionary past; could they be reactivated to expand the range of tastes and smells we perceive? What about changing people's character to make them more empathetic? A few authors have proposed a Genetic Virtue Project, which would use genetic engineering to improve our moral characters (more on that later).[2] More realistic technically are suggestions to modify a child's genome to increase their physical height or to choose a particular eye colour. There is undoubtedly a genetic component to intelligence, though this involves the subtle and complex interplay between hundreds of genes. Will it ever be possible to edit an embryo's genome to maximize a child's capacity for success? What about improved memory function, the need for less sleep or extended life expectancy?

Suggestions such as these aim at 'improvements' to the human condition, though even the most ambitious ones would not turn humans into beings of an entirely new kind. However, so-called transhumanists, such as the popular science writers Ray Kurzweil, Nicholas Bostrom or Yuval Noah Harari, propose making fundamental changes to human biology, speaking in almost theological terms of a new human species (e.g. Harari's '*Homo Deus*') endowed with god-like features (in contrast to *Homo sapiens*). Transhumanism may make for fascinating reading and speculation, but in this chapter we will discuss the less ambitious enhancements that modify a person's biology without transforming it entirely. These more modest proposals are technically realistic, yet can be ethically just as challenging.

In the area of gene doping, safety is certainly an important consideration. In the athletic enhancement described above, more red blood cells will make the blood thicker, increasing the risk of heart attack, stroke and blood clots. Several professional cyclists have died in their sleep after engaging in blood doping. Proponents suggest that gene doping should only be allowed up to safe levels. However, that is not possible to define, as different people react to the same interventions in different ways. It is also not clear that genetically modified athletes or their coaches would be content with the permitted enhancements alone, and could take further drugs to boost their chances of success. The skier Mäntyranta took amphetamines, which was legal at the time, in addition to his increased EPO levels. Since some athletes already accept the severe health risks that accompany illegal doping, the legalization of the practice or the creation of novel enhancements is unlikely to make people more honest or make the competition fairer. If, however, some athletes engage in doping and some competitors don't, or use it less intensely, doping will clearly reduce the excellence that is required for winning. None of that even takes into account the damage that would occur at amateur levels of sport, where the controls are less stringent and hobby athletes might be increasingly tempted to resort to dangerous technical 'fixes' in order to imitate their professional idols. Perhaps enhancements do not provide us with a bounty of good things like the cornucopia of the legends – do they rather resemble Pandora's box, offering difficulties we should rather avoid?

Obviously, safety must play an important role in any discussions about enhancement. However, we still need to think about the moral implications of any enhancements, even if the procedures could be made perfectly safe. In this chapter we will first discuss the technical feasibility of some enhancements, before considering their moral and ethical implications. There are both secular and theological ethicists who are in favour of enhancements. For example, the theologian Ted Peters argues that enhancements do not change the fact that humans are created in the image of God, so any enhancements that do not make dramatic changes to the human person could be legitimate.[3] Proponents of enhancements argue that they merely continue the pursuit of physical well-being, of which therapy is just one part. Critics reply that the continual striving for optimization is wrong and that issues of social justice and respect for what is natural require greater attention. They assert that enhancements may be inherently wrong. By contrast, one question we are keen to discuss is whether well-being is indeed likely to improve with enhancements. Enhancements are often described as interventions that help a child to be 'better than well'. Is that realistic even if the technology itself works? Does faster, brighter and stronger necessarily mean better?

Finally, an important question is whether we can distinguish between physical problems that require *medical* attention and enhancements that might be practised on *healthy* people. Famously, the World Health Organization once defined health as 'a state of complete physical, mental and social well-being and not merely the absence of disease or infirmity'. This has often been critiqued as unattainable, so the physical definition of health may not be so obvious.

Enhancements: the technical side

So just how realistic are these scenarios? Although genome-editing procedures for boosting athletic ability may be dangerous, they may be able to achieve the goal of increasing endurance. Cosmetic changes such as choosing a child's eye colour are technically possible and less risky. Although 16 different genes may be responsible for eye colour, most cases are determined by only two of these, located on chromosome 15.

What about modifying more complex traits, such as intelligence or height? Characteristics like these involve a combination of many small genetic differences, each of which alone has an imperceptible effect. Height is affected by at least 697 variations at 400 locations, while intelligence is influenced by hundreds of genes. Making a significant change to optimize these characteristics will require multiple genome-editing events, and we have no way of knowing whether these will have unintended effects on other traits that share some of the same genes.

The question of which parts of the genome to optimize is not straightforward scientifically. When researchers try to identify what genes have an impact on complex traits like physical height or academic performance, they compare the genomes of large cohorts (containing over a hundred thousand individuals) and examine the exact DNA sequence at hundreds of positions. Using computer programmes, they look at which DNA sequences at each position go together, or correlate, with a particular trait of the person. Each position alone has a very small, even imperceptible effect, but correlations begin to emerge from examining the different combinations, producing something called a polygenic score that indicates some of the genetic effects on the trait's variance.

However, this analysis does not capture the many ways that the genome is modified naturally by the addition of chemical markers to the DNA in individual cells, which regulate the genome. This is the subject of epigenetics. This emerging field studies how very similar genetic sequences in different people can have different effects. These markers don't change the DNA sequence itself, and their addition depends on many factors, including environment, nutrition or stress. At the moment our understanding of these processes is at a very early stage.

To explore the genetic contribution to intelligence, some scientists study the educational attainment of many people and analyse their genome. However, academic achievement is not simply an indicator of high intelligence, and intelligence is also due to non-genetic factors, such as industry and social support. There have been many geniuses like Alan Turing and Winston Churchill who did not do well in many academic disciplines, discovering their unique talents perhaps even despite formal academic requirements. For everyone who is admitted to university, there are an unknown number of intelligent people who are not. Further, there are so many genes involved that will each inevitably contribute to other traits as well; so, for example a gene that has a small effect on IQ might, for example, also contribute to male pattern baldness, short-sightedness or any other number of traits. In summary, it is not clear if improving intelligence with genome editing will be feasible. It makes more sense to invest in good education programmes than to seek a genetic enhancement of intelligence.

Even if we could influence intelligence by genome editing, we need to ask whether this is ethically appropriate. Parents might do this in order to give their offspring the best chances in life. But it is not guaranteed that educational achievement or high IQ make for a fulfilled and satisfied life, and any such assumption may be wrong. Some research even suggests that a high IQ is associated with a higher risk of mood and anxiety disorders.[4] So our common assumptions about what contributes to a more fulfilled life may be wrong.

Another prominent dream of potential enhancement is the hope of an increase in our lifespan. Life without death is the stuff of the ancient dreams of humanity. In the Sumerian epic of Gilgamesh (c.1500 BCE), the hero of the story seeks the secret of eternal life. In a story from the Bible, God banishes Adam and Eve from the Garden of Eden, declaring that they 'must not be allowed to reach out and take also from the tree of life and eat, and live for ever' (Gen. 3.22, NIV). In contrast, other places in the Old Testament imagine a time when 'No more shall there be in it an infant that lives but a few days, or an old person who does not live out a lifetime; for one who dies at a hundred years will be considered a youth, and one who falls short of a hundred will be considered accursed' (Isa. 65.20). Is genome editing perhaps one way to get there?

Several investigators have invested into the possibility of radical life extension for humans. The venture capitalist Peter Thiel, Amazon CEO Jeff Bezos and Google's Calico longevity lab spend a lot of money in efforts to extend life far beyond the 100th birthday. The US National Academy of Medicine also funds projects that aspire to postpone death for a long time.[5]

Ageing, the process they try to postpone, can scientifically be defined as 'the time-related deterioration of the physiological functions necessary for survival and fertility'. Although there are many age-related diseases, such as cancer, heart disease and dementia, ageing itself seems to be part of a normal healthy life.

Why do we die in old age anyway, and what mechanisms are responsible for ageing? One reason is that, with the exception of germ and stem cells, most of the cells in our bodies can only divide about forty times (something called the Hayflick limit). This happens because small sections of DNA are lost from the ends of our chromosomes every time that they divide. The ends of chromosomes are made up of repeating sequences called telomeres, which function like the plastic at the ends of shoelaces and protect the rest of the DNA. These get shorter every time a cell divides, until they are critically reduced. At that point, the cell enters a dormant state called senescence. In contrast, we know there is an enzyme called telomerase, which restores telomere length, from stem cells. So would reactivating telomerase in other cells protect them from some aspects of the ageing process, perhaps even making them immortal? This might seem to be a promising approach, though a major problem is that this enzyme is also reactivated in most cancer cells, which can also divide indefinitely. So reactivating telomerase might increase the likelihood of developing cancer. Moreover, the decay of telomeres is not the only cause of ageing. Cellular DNA is continually being damaged by things that we can't avoid, from our diet and from our environment. Even the oxygen we breathe damages DNA and generates random mutations. Fortunately, our cells contain a lot of enzymes that continually check and correct these errors, though a small number sneak through and accumulate over time. Increasing the activity of these repair enzymes might delay the ageing process.

In fact the real obstacles are such that even CRISPR might not place drastic life extension within reach, as we would need to tackle not just the ageing process itself but the diseases of old age. For example, the older we get, the more likely we are to suffer from cancer. Cancer would probably be one of the greatest problems in a dramatic extension of human life. Ironically, we are victims of our own success, as the older we get, thanks to improvements in medicine, the more likely we are to develop cancer. Of course, lifestyle factors like smoking, drinking, exercise, dietary habits and perhaps even sleep affect cancer risk. In addition to the age-related cancer risk, the incidence of other diseases increases with age, such as arteriosclerosis, heart disease, stroke and various forms of dementia.

This is the backdrop against which some scientists have suggested that even in the best circumstances, humans will not be able to live longer than about 120 years. The world's oldest people have increasingly come close to that mark, but a French lady named Jeanne Calment, who died aged 122 in 1997, is the only person verified to have lived beyond their 120th birthday.[6] Assuming that about 125 years is indeed the maximum lifespan, is it 'fixed and subject to natural constraints' that no technology can conquer? Or is that the maximum, cancer notwithstanding, because we have not yet deployed genome editing properly?

Enhancements: the moral question

The scientific details of genetic enhancements apart, why do many people have a negative view of non-medical enhancements? Is this reserve justified? And why are others so strongly attracted to enhancements? Is there a simple ethical line that separates legitimate medical interventions from dubious non-medical enhancements? How do we define illness, health and medicine, and distinguish these from the increase in physical ability that enhancements are seeking?

Many people think negatively about enhancements because they perceive them as unnatural. It is not uncommon among Christians to hear the view that enhancements must be inherently wrong as they attempt to improve on God's creation, which has already been declared as 'good' (Gen. 1). While we are critical of enhancements, this is not because we regard them as unnatural or deviating from God's good creation. Life today is already 'unnatural' compared to what it was like two thousand, five hundred years or even one hundred years ago. Literacy can change people's brain structure. One of the better-known examples comes from London taxi drivers, who memorize a plethora of geographic information and have an enhanced region of the brain known as the hippocampus. Life expectancy has dramatically increased; many previously fatal diseases can be cured, and we have numerous mechanical devices that support our failing faculties, including prosthetics and spectacles. Our human genetic diversity has increased since the days when most people didn't travel far and married others who lived in the same community or one nearby. Indeed, it has been said that the invention of the bicycle decreased the incidence of inbred genetic diseases!

In the Bible, God declared that creation is 'good' (Gen. 1). Note that the word 'good' here does not imply perfection, or even any moral quality, but simply that it is fit for purpose – it works and does what it is supposed to. Nature includes the fruitful biosphere in which it is possible to lead a fulfilling life. It does not mean that nature as made by God is idyllic or must remain as it is. Indeed, humans are commanded to 'subdue' the earth (Gen. 1.28). In other words, there was work to be done, to look after God's good creation and fend off existing forces of chaos that threaten creation. When Revelation 21 expresses a hope for a new heaven and a new earth, the profound transformation of life is expressed with the idea that two factors symbolizing forces of chaos, the sea and the night, will be no more. Nature as we find it is not sacrosanct, and human interventions in nature are not taboo merely because they change something that has been arranged differently in God's creation. Of course, not every human intervention in nature is good, and many turn out to be harmful. Yet we are not 'playing God' simply by modifying genes.

Nevertheless, we do seem to have powerful and healthy concerns for 'naturalness'. Here a common sentiment is that 'mother nature knows best' or that we 'must not play God'. These are convenient, popular shortcuts for making moral decisions about complicated issues, without the need to explore the details of empirical and ethical analysis. The words that are sometimes used in the media are constructed to engender feelings of revulsion or fear ('Frankenfood', 'Frankenmice' etc.). Even the term genetic *engineering* can seem cold, calculated and mechanical. However, humans have always been modifying nature, whether fighting disease, cultivating crops or managing the environment in which we live. Open-heart surgery (and especially heart transplantation) is just one example of medical procedures that we take for granted today but were first seen as breaking a taboo.

Nature has played a prominent role in traditional moral philosophy. In this view, nature is seen as the sum of the conditions under which human life is most likely to flourish and to be fulfilling. Order in nature carries moral weight primarily because it allows for activities that are excellent and inherently valuable. In acting justly and courageously, for example, we affirm human nature by going beyond mediocrity and practising the best possibilities of human nature. Actions that we admire often press against our natural human limitations, from the 'inside', as it were. We admire Usain Bolt precisely because his speed is exceptional and reminds us of what most of us cannot do. It would therefore be inappropriate to want to transform human ability fundamentally so as to do away with such limitations.

If natural disasters or disease prevent human fulfilment – if they disfigure human nature – then an intervention of some kind has traditionally been considered legitimate. However, there is an ambiguity here. The tradition that views naturalness as normative has considered changes *within* human nature partly as legitimate and partly not. However, it has not entertained the possibility of changes to the fundamental principles of human nature, and here one would wonder why they should be ruled out in principle. Our argument against doping in sports was that it is unsafe and compromises the excellence in sports by privileging some athletes over others, but not that it is unnatural. If an imaginary modification of human nature significantly increased everyone's running speed and did this safely, there would still be meaningful competition and excellence, so it may not be persuasive to assert naturalness as a norm against that. Of course, the question remains why one would want to do that in the first place, as competition and excellence are significant possibilities already.

Before we explore further some of the critical approaches to enhancements, we should also consider arguments in their favour. Society already accepts non-genetic enhancements, such as drinking coffee as a chemical stimulant, enabling us to be alert. Are genetic enhancements just a variation of that? The stimulant action of caffeine is not without its downsides. Caffeine-induced sleep deprivation causes accidents, and excessive consumption has been suggested to increase the risk of certain cancers.[7] There are trade-offs in its use, but since coffee is readily available to everyone, its slight benefits or disadvantages are equally distributed across society, so at least it does not raise questions of social justice.

Widespread coffee consumption may have negative *consequences*, but there does not seem to be anything *inherently* wrong in its use. Of course, all innovations have unintended consequences, though these can be minimized by regulations and monitoring instead of absolute prohibition. Electricity, cars and mobile phones are just a few examples of useful technologies that have some negative consequences. So new technologies might still be justified, although foreseeable problems need to be reined in with new rules.

Advocates of genetic enhancements point to the fact that expensive tutoring and an elite education are ways in which society already allows some 'enhancements'. Some people pay significant amounts of money to give their children an educational advantage. So why not allow for other forms of enhancement as well? The same might be said about sports teams that can afford specialized athletic equipment and exclusive training methods. Although the competitors rely only on their natural human talents and conventional training methods, money can legally buy athletic enhancements of a more conventional kind.

However, just because we tolerate one form of inequality does not mean we should introduce others. Many countries tolerate large differences in the availability of educational opportunities. Underprivileged people are less likely to have access to elite education that improves their social situation, and social mobility is diminished. Genetic enhancements could exaggerate those inequalities that are already inherent in the educational system, with those who invest in their own or their children's elite education using genetic enhancements to cement their advantage.

Genome editing could enable many to maximize their capacities, but it will also place many at the margins. In the area of life extension, there are presently large differences in life expectancy across the world. This is around 55 years in some economically less advanced countries, compared with over 80 years in many nations. There are often disparities in different regions within a country. Considering this huge gap in life expectancy, is it justifiable to work towards life extension for those who are already advantaged? It seems impossible for an entire population to access life extension equally, but is that a justifiable reason to block access entirely? Are greater biotechnological opportunities for a minority *necessarily* unfair? The philosopher John Harris notes that 'If immortality or increased life expectancy is a good, it is doubtful ethics to deny palpable goods to some people because we cannot provide them for all.'[8] If life extension is banned based on the principle of equality, then individuals who could have accessed the treatment will be denied a number of additional years of life.

If people pay privately for better medical care, is the resulting difference in health necessarily unjust, as long as good medical care remains accessible to others? Enhancement proponents comment that a strong insistence on equality appears as mere spitefulness that attempts to block benefits for others. On the other hand, the emergence of a biotechnological underclass does not seem right. The question is to what extent benefits for some imply a disadvantage for others. Can there be circumstances in which greater opportunity for some can be a good thing, perhaps even lifting everybody else's chances as well? In the UK, the Nuffield Council on Bioethics has stated that heritable genome-editing interventions are ethical if carried out in accordance with the principles of social justice and solidarity.[9]

While it can be a problem to deny goods to some in the name of equality, equality is by no means simply an empty and abstract notion. Genome editing could lead to greater wealth disparity, with the unedited 'have-nots' disadvantaged, compared to the 'haves' whose parents could afford the luxury of editing theirs. Increases in social inequality tend to diminish social participation and reduce political or economic opportunities for underprivileged people. In these circumstances, entrenched inequality seems unfair, and that would be a genuine problem with many enhancements. With tight regulations, perhaps genome editing would not further exacerbate societal inequality, as some suggest. However, in a world marred by racism and sexism, how can it be ethical to bring yet other sources of discrimination and inequality into society? Altering one's genes may lead to harmful social inequality and the possible development of a genetic caste system.

Are enhancements worth having?

Enhancements may seem tantalizing, while on closer inspection they may not be worth having after all. We need to shift the debate away from discussing enhancements only in terms of basic principles and norms to also include a realistic look at their consequences. Such an approach will draw our attention to the deeper questions of what it means to be human.

Many people would like the idea of having a genetically enhanced memory – but would this really be helpful? The ability to forget or to ignore is important and goes much deeper than a mere question of having a certain amount of mental 'storage space'. Our limited memories allow us to distinguish between the important and the trivial, rather than cluttering our minds with inconsequential details that could distract from the significant. Forgetting is also about distinguishing and evaluating, as the psychologist William James observed: 'Selection is the very keel on which our mental ship is built', he argued. 'If we remembered everything, we should on most occasions be as ill off as if we remembered nothing.'[10]

Forgetting also has an important social and moral function. In the Bible, the Old Testament even portrays God as forgetting – not because of inadequate memory but because of God's mercy. God is said to 'not remember your sins', as God 'blots out your transgressions' (Isa. 43.25). 'For I am about to create new heavens and a new earth; the former things shall not be remembered or come to mind' (65.17). Of course, remembering can be morally important: if one person mistreated another, it may be wrong to forget the injustice without proper reconciliation. Even if people have reconciled, remembering the conflict can help prevent new injustices. However, forgetting the particular sting of a wrong is an important part of forgiveness. There is a reason that the proverbial expression combines the two aspects: 'to forgive and forget'.

In another suggestion, ethicists have asked whether you would like to require less sleep, while still being well rested. Scientists have recently started to identify some of the genetic factors that affect our need for sleep and why some people seem to need less than others. If genetic modifications could safely allow people to feel rested with less sleep, then who would not be eager to gain an extra hour of free time each day? What would happen if we did? The idea might seem tempting for some people, although parents who would modify their child this way might be in for even more sleepless nights with their extra-wakeful child (as explored in Peter James' thriller, *Perfect People*). Once a critical number of people require less sleep, the extra waking time would probably be used to increase one's standard of living and so to do yet more work. As one's colleagues would have more time to send work emails, we would not engage in more leisure. That is at least what happened when labour efficiency increased greatly during the second half of the twentieth century. It is also doubtful that, above a certain threshold, an increased standard of living makes people any happier. Indeed, as gross household incomes doubled in the second half of the twentieth century, the percentage of people who reported feeling very happy remained constant.[11] Once a significant number of people gain an extra hour of wakefulness, it seems that society would merely switch the treadmill to a faster pace.

Related to this topic is the issue of radical life extension, which requires further ethical reflection in spite of the technical difficulties. 'I am grieved by the transitoriness of things', the philosopher Friedrich Nietzsche lamented in a letter to a friend. If we *could*, should we live much longer? But what sort of extension would we desire? Merely extending the time spent in old age does not seem satisfactory, as this will only increase the incidents of age-related disorders and put an even greater financial burden on care for the elderly. At what age would we wish to pause the process of ageing? In our twenties and thirties, at the height of our reproductive fitness, or later in life, in our fifties and sixties? Indeed, could the symptoms of old age be reversed? These may seem like unlikely and far-fetched developments, but they provoke us to think about the problems.

If we radically extend our life's period of health and vigour, it is plausible that we would become less aware of our vulnerability, helplessness and frailty. As one ethicist argues,[12] the conditions of infancy and old age would be further away from us than they are now, and so we would become less compassionate, as we are less likely to see the vulnerabilities of others. In *Gulliver's Travels*, Jonathan Swift explored yet a different problem. The Struldbrugs live for ever. However, the nation of Luggnagg prohibited them from holding private property after their 80th birthday, fearing that there could be a harmful concentration of resources in the hands of just a few people. The Struldbrugs were not exempt from the infirmities of old age. Let's assume we could take care of that with genome editing. Yet the Struldbrugs also have the problem that they do not remember recent events but are firmly moored in the experiences of early life.

As we age, and society changes around us, we risk being trapped in our memories and practices of the past. A radically extended lifespan would turn us into 'walking anachronisms who have outlived themselves', in the words of the philosopher

Hans Jonas.[13] Times change, and we have to change with them. With radically extended lifespans, we and our contemporaries will change in different ways, and over many years the differences will accumulate. The longer we live, the more will our previous moral formation diverge from current cultural practices, and the less we will feel at home in the new social and moral context. In addition, we might improve our short-term memory. However, we have already seen that to remember things does not simply mean to store things but to evaluate and to appreciate them. Since to appreciate and to remember means to select some things over others, an infinite memory within a radically extended life would run the danger of appreciating all sorts of things, but nothing in particular. We might end up with personalities that are incoherent or trivial.

To be sure, 'Old age sure ain't for sissies', as Bette Davis said. Our bodies become frailer and our situation in life changes. Ageing takes courage, which is a virtue that must be practised. Even those with radical life extension will need it, but we do not become more courageous by putting ever greater efforts into fleeing from the challenge. Radical life extension becomes even more questionable if ultimate redemption does not exclude biological death, but embraces it, as the Christian faith holds. Only in the face of our biological death does God defeat death and give us eternal life. God's redemptive future, which 'no eye has seen, nor ear heard, nor the human heart conceived' (1 Cor. 2.9), also promises to overcome injustice, sin and suffering. That is not achieved by extending our present lives but by a radical transformation in the life of the world to come. Escaping from biological death would attempt to evade God's liberation from that burden. Life is good and valuable, and we hope for a life-span similar to the proverbial 'threescore years and ten' (Ps. 90.10, KJV), or perhaps even fivescore. Beyond that range, however, any gain in extra years would seem to be of marginal value, given that biological life will still continue to be lived in that shadow.

Many of the proposed enhancements are for the sole advantage of the individual. However, there have also been suggestions that technology could make people less focused on themselves and instead more caring and altruistic. For example, germline editing might be used to increase the number of receptors for the hormone oxytocin, which is involved when we feel empathy for someone else. It has been shown that voles can be modified genetically to change the social behaviour of laboratory animals, although these changes were not seen when the animals were released into their natural habitat.[14] An anthropologist has observed that we increasingly speak of empathy and suffering rather than of justice and injustice.[15] However, would the parable of the Good Samaritan have ended differently if the priest and Levite who passed by the injured man without helping had had genetically increased levels of empathy?

Perhaps the argument for an enhancement would let those who failed to help the robbed and beaten man off the hook too easily! Biologically, we are highly empathic creatures *already*. However, we often fail to translate empathic feelings into principled action, and in addition, compassionate feelings may steer us in a variety of directions. Loving one's neighbour is not simply a matter of inner feelings but of conscious decision and practice. Our moral car needs the energy of our emotional engine for motivation, and empathy can give us clues where to go. Yet when choosing the right route, decisions made on the spur of the moment can lead us astray. Instead of more empathy, what we require is more reflection on justice as well as practice. Justice is to be compassionate towards victims out of the conviction that it is the *right* thing to do. We need to select which urges to act on and calibrate our response. In this sense, justice is at the heart of morality. Conflicts are often more complicated than the black-and-white scenario of an evil perpetrator preying on an innocent victim. Think of a judge sentencing a defendant to prison – a relative of the victim of the crime and a relative of the defendant may feel empathy with different people! The first person with whom we may have compassion is not necessarily the party that is morally right. More empathy is not a shortcut to a better personal morality. To enhance morality, an improvement in moral reflection and the formation of good habits would be required, but genome editing cannot help us with that.

What's morally wrong with enhancements

So in discussing potential enhancements, the question is whether we even know what we are wishing for. Not only are we finite creatures, we are also creatures with limitations who adapt and find meaning in life, even as a direct result of contending with these limitations. We need to challenge the assumption that greater physical and mental capabilities will produce more fulfilling lives. The logical extension of such a mindset is that anyone with capabilities at the lower end of the spectrum is deficient in some way.

Traits perceived as socially less desirable would increasingly be 'deselected' as people adjust their children's or their own genome. The idea of enhancements largely follows the idea that more is better, which often appears deceptively true. If, for example, an impairment reduces mobility or causes people to feel fatigued very quickly, it seems reasonable to claim that a person's quality of life will be affected. Yet if we extrapolate from that to non-medical enhancements, suggesting that increased functioning will improve well-being beyond what is normal, we will encounter the law of diminishing returns. Surveys among people with disabilities confirm that they typically adapt well to their situation in life. For example, the scholar Tom Shakespeare reflects on how well-being improves when people embrace the way that they defy common beauty ideals: 'For disabled people, we are always different from the norm. So we have nothing to hide, and no reason to feel embarrassed by undressing. Paradoxically, we may be happier with our seriously flawed bodies than others are with their minor defects.'[16]

Not only do people adapt well to perceived disadvantages, experience also shows that some of the things that we think of as highly desirable do not have a lasting effect on our well-being. Winning the lottery gives people a positive feeling only for a very limited period of time. Nevertheless, it seems to us that the grass is always greener on the other side, that having more of this or that would make us happier. In one study, researchers asked Americans whether life is better in

California, compared to other US states. Everybody agreed – except Californians.[17]

One potential harm in prohibiting enhancements is that individual, autonomous choice will be more limited. In Western society, our intuitive sense is that more choice must necessarily be better. According to this view, the only justification for limiting choice is to avert harm. However, psychologists have questioned whether more choice results in greater happiness.[18] Experiments suggest that having made a decision for some product on the spot, people enjoy more what they choose, compared to those who were asked to revise or confirm their initial choice later on.

This confirms our argument that enhancements are unlikely to fulfil the hopes people place in them. Society will not be denying people a genuine good if we do not permit genetic enhancements.

Just as importantly, we now turn to the argument that enhancements would create harm, even if procedures were safe from a technical perspective. Editing the genome to increase particular characteristics because 'more is better' may denigrate people living with genetic diseases or disabilities who live a good life and do not see themselves as born with a 'genetic error'. Here enhancement proponents typically respond that we can have it both ways and one thing does not rule out another: we can respect others and the lives they lead while also pursuing a different kind of life for ourselves and our children. To be sure, a pregnant woman who maintains a healthy diet to avoid harming her baby does not thereby belittle people with chronic illness. However, to seek enhancements beyond such common health concerns means going to extraordinary lengths, and that is a different matter. If society allows for non-medical enhancements, we would be accepting, despite good reasons to the contrary, that more functioning is better. That would reinforce the idea that bodies with greater or lesser capabilities are not part of a legitimate diversity but rank higher or lower on a scale of value.

Besides the potentially harmful effects of genetic enhancements on social values, they are also questionable on other grounds. This is because the genetic technology works most effectively when used on embryos, rather than children or adults. This raises the question of whether parents can legitimately choose enhancements for their children. Under what circumstances are parental choices on behalf of their child legitimate, and when are they not?

The philosopher Joel Feinberg asserted 'a child's right to an open future',[19] which seems to speak against enhancing a child. Jürgen Habermas insists similarly on the principle of human self-determination, as the child will have had no part in the genetic decisions that affect their later life. Michael Sandel argues against embryo enhancements from an understanding of the 'gift-character' of life, which implies a fundamental open-endedness in the child's future. He contrasts this with an overly ambitious trend towards optimization of the child's potential. However, others contradict this and suggest that genome editing is merely another means of improving a child's future prospects. Parents rightly act as 'surrogate decision-makers' for their children, for example in health-care matters. They already make plenty of other choices on behalf of an unborn child and will continue to do so for several years. Further, genes do not undermine the child's agency in a deterministic way. An athletic enhancement does not by itself turn the child into an athlete. Children will still control the outcome of the enhancement when deciding what, if any, sports to pursue and how intensely to practise. Others suggest that when considering embryos, only those enhancements that anybody would desire should be regarded as legitimate. Some even ask, if germline enhancements can improve a child's future, is it not the parents' ethical responsibility to make use of them?

On the whole, however, these arguments are rather idealistic and ignore parental expectations and pressures, such as when 'helicopter parents' and 'tiger mothers' exercise undue pressure in managing the lives of their 'overscheduled children'. The harmful mental health effects of these widespread phenomena are well documented. Parents who choose an athletic enhancement for their child are less likely to respect the free choice of the 'enhanced' child later on. What if an athletically enhanced child chooses to be a bookworm?

Until the advent of precise biomedical technologies, parents knew nothing about their offspring until they were born. They may have had high aspirations for what the child might become, but the future was unknown. Without the biotechnological means to modify offspring, children could only be loved and cherished for what they were, rather than measured against any supposed chosen future. Parents have always exercised some influence on their children to take one path in life rather than another, but at least such pressure was exercised on the basis of hopes and wishes, rather than influencing the outcome with additional biotechnological means. We appreciate our children as gifts and accept them for what they are; as they develop not as products of our design or will. Parental love should not depend on the talents or attributes of the child. We may choose our friends, hoping they will complement our own personalities, but we do not choose our children. Their characters and qualities elude our control, no matter what good influences are provided by their parents. As Francis Collins has said: 'The application of germline manipulation would change our view of the value of human life. If genomes are being altered to suit parents' preferences, do children become more like commodities than precious gifts?'[20]

The Christian ethicist Gerald McKenny has argued that enhancements are not necessary to enjoy the life with God for which we were created. Humanity is created in the image of God, and we are able to live up to this calling without enhancements. While this leads McKenny to be cautious about enhancements, Ted Peters draws a different conclusion, though from a similar appeal to our creation in the image of God: as minor enhancements do not change the theological fundamentals of human nature, they are probably a good thing. However, these positions fail to ask why one would want enhancements in the first place. Some people will answer that the reason lies in a hope for some improvements to life's potential, but for all we know, that is an illusion that could have harmful consequences.

How do we distinguish medical interventions from enhancements?

Since there is no reason to think enhancements will improve life in the way advocates hope, our suggestion is that genome editing should be limited to medical and therapeutic procedures. Life extension may be a special case where the boundary between medicine and enhancement becomes blurred, although it is not clear what the real benefit of a drastic life extension would be. However, advocates of enhancements suggest that restricting genome editing to medical purposes is bound to fail, as we cannot define health and distinguish it from enhancements.

Some people understand health as a state that is typical of the species. Following this definition, one might define a new vaccine as an enhancement, rather than a medical procedure, since it generates a new ability to defend against a particular pathogen. However, inoculated people do not have any new capacities – their bodies function in exactly the same way as those of people who have never encountered the pathogen. The vaccinated person might not even know if the vaccine actually has made a difference. In any case, this is particular to the individual and does not create any changes that will be passed from one generation to another. Further, defining health in terms of what is statistically common for the species has little to do with the crucial point: people consult a physician not because they want to conform to a statistical norm but because they want to feel better.

A medical diagnosis of an illness recognizes that there has been a disruption in an organ's proper function. This is what causes discomfort and prompts people to consult a physician. Here an organ's function is defined as the role it plays in the preservation and reproduction of the individual. There is little doubt about the proper function of the heart and the lungs, and their malfunction in heart and lung diseases has profound implications for living a healthy life.

Less life-threatening are things like near-sightedness, a less than optimal functioning of the eye, which will impede activities in our physical and social environments. Prescription glasses improve eyesight more or less, rather than restoring the entire function of the eye. If, however, medical interventions improve organ function, can we still distinguish clearly between medical interventions and enhancements? After all, enhancements may be done to achieve gradual improvements in existing functions as well. However, would an improvement in visual acuity even beyond 20/20 vision be a genuine improvement, comparable to the helpful effect of conventional spectacles? That is unlikely. With eagle eyes, we might see many details that we might otherwise miss, but as a trade-off, we might be less quick to see the bigger picture. When driving a car, being distracted by minute details that have nothing to do with traffic would be harmful, for example. We should not look at individual capacities in isolation from the wider body, or even the body's functioning in the environment.

This understanding of health has roots in the work of Plato, who highlighted that the medical arts require knowledge of the entire body. That is not merely because an ailment can occur in any part of the body, but because injury to one part has consequences for other parts. To define health in terms of the proper function of organs is to speak of their role in the wider context of the body. Indeed, the philosopher Hans-Georg Gadamer understood health as a systemic property that requires a balance between various aspects of the body.[21]

If we are to restrict the use of genetic modifications only to medical uses, based on a definition that emphasizes an organ's function in the body, does that mean that nature itself is the moral arbiter? If a desired function is merely what happens naturally, whatever that might be, then does this make medicine, which preserves or restores proper functioning, into something that is inherently conservative? No, for two reasons. First, disease itself is natural, so medicine is already used to correct or repair nature. Second, well-being is more important than what happens 'naturally'. The primary reason why we wish to restore an organ's function is not because the function is natural or unnatural but because that improves well-being. We suggest that, apart from the medical correction of a physical deviation from a natural function, it will be very difficult to use genome editing to improve functioning, beyond our natural boundaries, in a way that also increases well-being. Nature has evolved over millions of years, by processes of trial and error, and that has resulted in whole systems that work remarkably well, within natural physical constraints. Attempts at enhancement are likely to make things worse.

We have discussed what medicine is because we contend that genome editing should be restricted to medical uses. This requires that we distinguish medicine from enhancements. Based on our criterion of organ function, any intervention that restores or preserves the function of an organ is medical. There are critics, however, who say that the distinction between medicine and enhancements is unworkable in principle. They may respond to our suggestion that our particular definition, which relies on an organ's function, is a subjective criterion, while they ask for an objective one. However, contrary to our common intuitions, it is almost impossible to describe this in strictly objective ways, and defining health and illness can never be a purely exact science. Indeed, scientists and philosophers even find it difficult to define life in exact terms and need to rely on the use of intuition. However, while there is an intuitive, subjective element for determining the function of an organ, such a judgement is open to empirical testing and critical discussion, even if it may not yield an unambiguous, purely objective picture. Nevertheless, it often takes subjective, intuitive judgement to make the best sense of empirical measurements. Although definitions of health are a matter of broad, public debate, the opinions of scientists and medical experts should be valued particularly highly in the discussion of what is a disease. That is not merely because they have solid, testable knowledge, but also because they have trained their intuitive judgement in medical practice and in discussion with other experts. An example of scientists needing more questioning by a critical public, however, was the widespread, false assumption that black people had a higher pain tolerance than whites. An interplay between medical expertise and wider critical debate in society is our best hope for better medical knowledge.

By contrast, if our definition of illness and health depended merely on personal judgements or individual preference, rather than our shared human experience, could we deny people the right to use genome editing for other purposes than the healing of disease? What seems like an enhancement to one person might be considered as a medical necessity for others. However, if medical standards are simply arbitrary, it would not be obvious for us to oblige people to pay for medical insurance or to use taxpayers' money for the upkeep of public hospitals, medical schools and nurse training programmes. In part, this is because we can't arbitrarily call anything a disorder, or an enhancement. Being left-handed used to be seen as wrong ('sinister'), but of course one hand is as good as the other. Neither do freckles impede the functioning of anything, though some might wish to remove them. In these and many other cases, the scientific emphasis on measurement and analysis can help prevent discrimination and undue medicalization. This illustrates that any intuitive judgement about functions must command support from medical experts. The fact that our medical intuitions have changed over the centuries is in part because science has increased our understanding of how things work, and new intuitions can make better sense out of new scientific findings. Another example is society's approach to mental illness. In previous centuries, society was quick to consign people to psychiatric institutions. The ways we treat mental illness have radically changed, yet that doesn't mean medical diagnosis can flexibly portray anything or nothing as a medical condition. This change in the treatment of mental conditions is progress, but it doesn't mean medical diagnosis lacks all objectivity.

Any definition of illness will always be both 'naturalistic' (as objective and impartial as possible) and 'constructivist' (value-based). Those who advocate radical human enhancement will often see these as rival ways of defining illness. But this dichotomy is artificial, as these definitions are not mutually exclusive and cannot be taken in isolation from each other. There are of course some cases in which the distinction between medical therapy and enhancement is not clear-cut. Vitiligo, or light skin spots, can increase the risk of skin cancer, and therefore might be considered as a potential medical condition. For others it causes social discomfort as our faces form an important part of the way that we interact with other people. Hirsutism, or very dense facial hair, deviates from common aesthetic standards, so it might affect social interactions. However, in both cases the problem really lies in society's aesthetic standards, not in the presence of light skin patches or hair growth.

The question of what modifications are necessary or permissible is one of determining which functions are important. For the boundary cases, a broader social debate, a solid medical discourse and tolerance in society will be essential. However, these boundary cases do not mean that there is a false distinction between illness and health or that this is merely a matter of individual preference. For example, if we could remove cellulite with a genome-editing treatment, that would surely be considered a cosmetic enhancement. However, the media typically portrays cellulite in medical terms, and talks about the 'symptoms' of the 'condition'. This discourse is of course strongly influenced by commercial interests. Nevertheless, cellulite treatments are cosmetic, not medical, as cellulite does not harm any of the body's functions.

Conclusions

The philosophical debate about genetic enhancements is fascinating, though we should be careful to distinguish between suggestions that are technically realistic and those that are mere flights of the imagination. The possible genetic enhancements that we have considered include those for greater athletic prowess, intelligence, memory, longevity, the ability to thrive on less sleep, and greater empathy.

A particular issue in embryonic genome enhancements is that parents would be imposing them on their children. This would be done with the intention of giving their offspring greater opportunities or success. However, parents would run a significant danger of treating children as commodities, and they will be more likely to put pressure on their enhanced children to conform to the enhancement that they have purchased.

Some people suggest that enhancements are inherently wrong, independently of consequences. On that question we find that there is little reason to reject enhancements specifically as attempts to 'play God'. However, there are problems with enhancements besides the fact that they may not result in any genuine benefit. The very motivation for pursuing enhancements is questionable and is based on the assumption that greater physical or mental capacity will make for a better life. The consumer mentality of 'more is better' is simplistic and in many cases turns out to be wrong. People with reduced physical or mental capabilities report high levels of life satisfaction, and there is little reason to assume that those who have been enhanced will experience any greater sense of fulfilment. The assumption that enhanced capabilities are better – and that those people with enhanced capacities are in some sense superior – contributes to a climate in society in which people with disabilities are increasingly stigmatized for their differences.

It therefore seems best to restrict genome editing to medical uses, a topic covered in Chapter 3. Medicine can be distinguished from enhancements as the preservation or restoration of the proper functions of bodily organs. Of course there will be debate over what the boundaries of 'proper functions' are, but there is a general consensus among medical experts that resonates with public discourse. Human well-being is paramount, involving medical perspectives on medicine and health.

Some might then ask whether we should support any *medical* efforts in genome modification. Indeed, this type of medical progress can be a two-edged sword, which can contribute to the stigma of those who remain impaired, while relieving suffering of others. Here we will need to find a balance between different outcomes. The treatment of persistent conditions such as severe pain can improve the quality of life, and the offer of medical genomic assistance may be one way of giving people greater control over their own lives. We clearly need to provide good medical care for as many people as possible, including the use of genome modification, while being aware that such efforts can increase the stigma for those for whom medicine does not help. We need to beware of turning genomic medicine into an all-out fight against all human limitations.

Notes

1 Jocelyn Kaiser, 'Genetic Modification could Protect Soldiers from Chemical Weapons', *Science*, 22 Jan. 2020, doi: 10.1126/science. abb0103.

2 Mark Walker, 'Enhancing Genetic Virtue: A Project for Twenty-first Century Humanity', *Politics and the Life Sciences* 28 (2009): 27–47; Ingmar Persson and Julian Savulescu, *Unfit for the Future? The Need for Moral Enhancement* (Oxford: Oxford University Press, 2012).

3 Ted Peters, 'Can we Enhance the Imago Dei?', in *Human Identity at the Intersection of Science, Technology and Religion*, Nancey C. Murphy and Christopher C. Knight, eds (Farnham: Ashgate, 2010), pp. 215–38.

4 David Z. Hambrick and Madeline Marquardt, 'Bad News for the Highly Intelligent', *Scientific American*, 5 Dec. 2017, https://tinyurl. com/y76yfz8g (accessed 24.9.2020).

5 Tad Friend, 'Silicon Valley's Quest to Live Forever', *The New Yorker*, 3 April 2017, https://tinyurl.com/life-forever-quest (accessed 24.9.2020).

6 Xiao Dong, Brandon Milholland and Jan Vijg, 'Evidence for a Limit to Human Lifespan', *Nature* 538 (2016): 257–59.

7 Matthew Walker, *Why We Sleep: The New Science of Sleep and Dreams* (London: Allen Lane, 2017).

8 John Harris, 'Immortal Ethics', *Annals of the New York Academy of Sciences* 1019 (2004): 527–34.

9 Nuffield Council on Bioethics, *Genome Editing and Human Reproduction: Social and Ethical Issues* (London: 2018) https://tinyurl. com/yy6pjgne (accessed 5.10.2020).

10 William James, *The Principles of Psychology* (New York: Dover, 1950), p. 680.

11 David G. Myers, 'The Funds, Friends, and Faith of Happy People', *American Psychologist* 55, no. 1 (2000): 61.

12 Gerald McKenny, *Biotechnology, Human Nature, and Christian Ethics* (Cambridge: Cambridge University Press, 2018), ch. 5.

13 Hans Jonas, 'The Burden and Blessing of Mortality', in Jonas, *Mortality and Morality: A Search for Good After Auschwitz*, ed. Lawrence Vogel (Evanston, IL: Northwestern University Press, 1996), pp. 87–98.

14 Denis Alexander, *Genes, Determinism and God* (Cambridge: Cambridge University Press, 2017), ch. 5.

15 Didier Fassin, *Humanitarian Reason: A Moral History of the Present* (Berkeley, CA: University of California Press, 2011), 'Introduction'.

16 Tom Shakespeare, 'Does Cosmetic Surgery really make People feel Better about their Bodies?' *BBC Magazine: A Point of View*, 22 Jan. 2016, https://tinyurl.com/y5eje8eu (accessed 24.9.2020).

17 Daniel Kahneman, *Thinking, Fast and Slow* (London: Penguin, 2012).

18 Daniel Gilbert, *Stumbling Upon Happiness* (New York: Vintage, 2007).

19 Joel Feinberg, 'The Child's Right to an Open Future', in *Whose Child? Children's Rights, Parental Authority, and State Power*, William Aiken and Hugh LaFollette, eds (Totowa: Rowman and Littlefield, 1980), pp. 124–53.

20 www.statnews.com/2015/11/30/gene-editing-crispr-germline (accessed 19.1.2021).

21 Hans-Georg Gadamer, *The Enigma of Health: The Art of Healing in a Scientific Age*, trans. J. Gaiger and N. Walker (Cambridge: Polity, 1996), p. 88.

5

Eugenics

The word 'eugenics' has been used to describe ways for 'improving the human gene pool', or at least for preventing it from 'degenerating' over time. Perceived improvements could include an increase in health, fitness or intelligence. However, the historical practice of eugenics has not only been riddled with flawed science but it has also been associated with practices that were often downright evil, even including those carried out by the Nazis in Hitler's Germany. The issue of eugenics has also been raised in the debate about genome editing. When one geneticist recently argued for the genetic modification of human embryos to prevent heritable disease, he acknowledged that this might raise 'fears relating to the use of the technology for eugenics'. Indeed, a campaign group called 'Stop Designer Babies' responded, arguing that 'It is vital that ordinary people stand up and say no to a future of eugenics.'[1] Are contemporary attempts at human genome modification similar to historic attempts to change human genetics? Are those who see a similarity being alarmist, or are there lessons that can be learned from past mistakes that should inform the ethical use of present-day genome modification?

In contrast to its critics, some have tried to reclaim the controversial term 'eugenics'. A few years before the discovery of CRISPR, the philosopher Nicholas Agar promoted genetic enhancements, which he described as 'liberal eugenics', a term that he understood in a positive sense. He argued that genetic enhancement technologies could be made available to ordinary citizens, while emphasizing that individuals must not be coerced into undergoing any genetic modifications and that the best interests of the child must be foremost.[2] The expression 'consumer eugenics' is sometimes used in a similar sense, although it can also imply a criticism, and the 'e-word' is often understood in pejorative, inflammatory ways. It has of course been associated with the extreme practices in Nazi Germany. But is it a fair criticism of genome editing to draw a parallel to Nazi eugenics?

One of the motivations for eugenic programmes in the early twentieth century in the West was the fear of biological 'degeneration'. Hitler shared these fears, and the Nazis were infamous for their radical eugenic programmes, through which they hoped to turn Germans into a pure breed of physically superior people in a militaristic race state. Although eugenics would later be associated with the Nazi ideology, which was extreme and violent, there were also well-intentioned progressives among the earlier proponents of eugenics, in pre-Nazi Germany and elsewhere. The true motivations of historical agents are difficult to evaluate, but many were attracted by anything that could make people stronger and more intelligent. These eugenicists argued that science could be used to guide social policies and to create a better society, though they readily accepted that people perceived as 'inferior' would have to give up some of their freedoms in order to protect wider society.

Present-day advocates of genome editing complain that they are associated with evil Nazi policies in a fallacy that they term 'reductio ad Hitler'. They suggest that the wide availability of genetic enhancement technologies should not be equated with earlier flawed eugenic practices, notably by the Nazis, merely because they share a commitment to healthier, fitter lives. They emphasize that their proposals should be carried out without any compulsion or coercion, by the state or anyone else, and that society must not discriminate against people with reduced physical or mental capabilities. In addition, these contemporary proponents of genetic enhancements are not motivated by a fear of large-scale genetic degeneration of human fitness as the eugenicists were 100 years ago, nor are they interested in elevating one racial or social group over another. A case for 'liberal eugenics' could point to the fact that all parents already seek good health and the best environment for their children, through good nutrition, education, health care and a loving family. In a sense, one might say that it is part of being a responsible parent to strive for a 'good birth' – which is part of the meaning of the word 'eugenics'. Is it such a large step to encourage parents to ensure that their children have good genes? Are we perhaps even morally obliged to ensure the genetic well-being of our future children? On the other hand, can the aims of the present potential genetic modifications be clearly distinguished from the morally impermissible eugenic programmes of the past? Contemporary advocates of genetic modifications also propose enhancements that are not passed on to future generations, such as those performed on individual adults, while traditional eugenics aimed at population-wide interventions that would have effects across generations. Our understanding of genes and how they affect different aspects of our bodies, minds and characters is also scientifically much more advanced than it was a century ago. Our knowledge of how genes interact with our environment offers many possibilities for intervening in – even optimizing – the process of reproduction. How ought we to use this new knowledge?

This chapter asks whether the analogy with eugenics is useful in evaluating the possible application of present-day biotechnological enhancements. Considerations of deep interventions in human biology do indeed make many people uneasy as we struggle to find ways to help us make sense of the new technologies and to grasp their moral dimensions. However, to equate all genome editing, which might be intended to alleviate disease in adults and which might not even be passed on to future generations, with the evils of previous eugenic programmes *is* surely inappropriate. On the other hand, historical inaccuracies have crept into the account of Nazi eugenics. Moreover, an exclusive focus on *Nazi* history, when discussing the 'new eugenics', can distract from other morally significant features of eugenics as it was practised in many different countries about a century ago.

In this chapter we will argue that genome editing should not be portrayed as a variation of eugenics, either as practised in Nazi Germany or in other countries. Nevertheless, a study of the history of eugenics makes a helpful contribution to the discussion about the uses and motivations for performing genome editing. For this reason, we will discuss what eugenics has looked like in various countries and how this history can shed light on the current bioethical debate.

A history of eugenics in the West

Although the term 'eugenics' wasn't introduced until the nineteenth century, some of its ideas have a very old history. The ancient Greek philosopher Plato suggested that human reproduction should be controlled by the state and, realizing that this would not be popular, devised a rigged lottery system to choose the 'right' partners. The aim was to improve the human race, though Plato realized it was flawed, as many people with high scores could still produce offspring in lower numbers.[3]

In classic antiquity, several cultures practised infanticide through the exposure of babies who were regarded as inferior or disabled. In parts of Roman society, impaired children were killed, and patriarchs were known to 'discard' infants. The Roman Law of the Twelve Tables from 450 BC not only allowed a father to kill a child but even required him to do so: 'A father shall immediately put to death a son who is a monster, or who has a form different from that of the human race.' Indeed, the practice of infanticide in the Roman Empire persisted until the advent of Christianity, which, in the words of a recent commentator, 'marked a turning point in late antiquity in its appreciation of human life as having intrinsic value'.[4]

The modern version of eugenics started with Darwin's cousin Francis Galton in the nineteenth century. Galton was interested in 'improving human stock' by employing a seemingly scientific management of mating to create 'better' humans. Fearing 'degeneration', he argued that 'weakly and incapable' people should be prevented or discouraged from having children. This kind of effort is often called 'negative eugenics'. On the whole, however, Galton was more concerned with promoting increases in intellectual and physical vigour by encouraging marriages between those who were deemed to be physically and mentally fit, which is often called 'positive eugenics'. Yet it was the more repressive negative eugenics that was to become prominent later on.[5]

Galton's eugenics is often mentioned in the same breath with another ideology of the time, so-called social Darwinism, of which Herbert Spencer was a leading advocate. Both Galton and Spencer took Darwin's evolutionary insights about the 'survival of the fittest' and applied these to human society. By arguing that 'a creature not energetic enough to maintain itself must die' (Spencer), they opposed the provision of humanitarian aid to every human being. They claimed that indiscriminate health care would be harmful to society by allowing the weak to survive and reproduce. While social Darwinists thought that family lineages with seemingly undesirable traits would eventually die out under the pressure of natural selection, eugenicists asserted that an extra repressive effort would be required to keep less intelligent and weaker people from diluting human fitness. Although some of Darwin's writings can appear to support social Darwinism, he was committed to humanitarian principles. He critiqued social Darwinism and defended the indiscriminate provision of health care to all classes and groups, which he saw as 'the noblest part of our nature'.

The scientific understanding of these early eugenicists was wrong in a number of ways. They believed that simple biological inheritance played a crucial role in determining human traits, such as intelligence and criminality, and idealized strength and competitiveness. These eugenicists used their erroneous beliefs about evolution and biology to predict or to defend behaviours and responsibilities. They did this despite the fact that biology, and evolution in particular, does not make claims about what is right and what is wrong. Evolution is merely a historical description of how living organisms have come to be as they are, and is necessarily silent on the moral imperative of defending vulnerable members of society. Or as the historian of science Steven Shapin wrote, 'Everybody knows that the prescriptive world of *ought* – the moral or the good – belongs to a different domain to the world of *is.*[6] Darwin's collaborator, Thomas Henry Huxley, even wrote that 'cosmic nature is no school of virtue, but the headquarters of the enemy of ethical nature.'[7] While Huxley was a vocal critic of religion, his moral views were informed by a strong emphasis from Judaism and Christianity that requires people to defend the weak and support the poor.

In the early twentieth century, societies for the promotion of eugenics were founded in Britain, the USA, Germany and other countries to address biological and political anxieties. In the USA, Scandinavia and Nazi Germany, laws allowed for the sterilization of people who were considered to be undesirable. At the International Eugenics Conferences, held in 1912 in London and later in New York, politicians, activists, scientists and philosophers considered that their hard-earned contemporary cultural and political achievements might be in danger. They thought that the people they branded as less intelligent or criminal tended to have more children than average and would therefore increasingly gain in number and influence. Of course, we do not know of any genetic determinants of criminality, lewdness or antisocial behaviour, but speculation about such biological inheritance made intuitive sense to many people at that time.

In the USA, strong and healthy families with several children were awarded eugenics prizes at local county fairs, in the same way that farmers are given awards for their healthiest and most vigorous animals, which are then used for selective breeding, excluding the less fit ones. A sign at rural fairs asked, 'How long are we Americans to be so careful for the pedigree of our pigs and chickens and cattle, and then leave the ancestry of our children to chance, or to "blind" sentiment?' Several groups, including Protestant pastors, joined the propagandistic effort. In order to reduce the number of children born to people with physical or mental illness, disabilities, or to criminals, eugenicists encouraged or forced sterilizations, especially of women who were deemed unfit, including the poor, mentally insane, 'feeble-minded' and drunkards. Over 30 American states legalized forced sterilizations between 1907 and the mid-1930s, and the Supreme Court decision *Buck v. Bell* (1927) declared the procedure legal. The Supreme Court judge Oliver Wendell Holmes, Jr famously declared that the public purse should not be unduly burdened and that 'three generations of imbeciles are enough'. The Nazis studied the US discussion very closely.

In California, the majority of the 20,000 forced sterilizations were done to Hispanic women, while in Virginia, African American women suffered the most from about 8,000 forced sterilizations. In total there were about 63,000 forced sterilizations in the USA.[8] Presumably these measures were not simply taken out of malice, but often with the intention of 'improving the human gene pool'. The eugenics laws were not removed from the books until the 1960s, though very few people were forcibly sterilized after the war. Although the policy has been overturned, the noxious, false assumption that ethnic groups differ in intelligence levels, due to their supposed genetic profiles, can still be found in print today. In 2007, the Nobel laureate James Watson provoked controversy when he stated that he was 'inherently gloomy about the prospect of Africa' because 'all our social policies are based on the fact that their intelligence is the same as ours, whereas all the testing says not really'.[9]

The eugenics effort was of course based on unproven or incorrect scientific theories, in which many activists mistakenly assumed that conditions such as 'pauperism', criminality and 'feeble-mindedness' were passed from parent to child and inherited, rather than arising from a combination of factors under conditions of social deprivation or insufficient education. An example of this way of thinking can be seen in the words of the eugenicist Charles Davenport: 'Though capital punishment is a crude method of grappling with the difficulty [of those with inferior genes], it is infinitely superior to that of training the feeble-minded and criminalistic and then letting them loose upon society and permitting them to perpetuate in their offspring these animal traits.'[10] If such an analysis were to be correct and if human nature were predominantly determined by our genes, then it would be logical to conclude that in order to change society, we need to change the gene pool. Taken to extremes, this form of genetic determinism views society's problems as arising from within the genomes of the individuals who make up society. The solution is therefore to restrict the freedom of these individuals or even kill them, not to challenge the existing social structures directly. As we will see below, this way of thinking led to the atrocities in Nazi Germany in a fanatical pursuit of fitness ideals. Yet the discussion also had a contrasting, equally wrong political counterpart. Against the genetic determinism that was implied by Western eugenics, Communists claimed that education and the environment were the dominant factors in biological nature, and that even plants and animals could be educated. The false dogma of the inheritance of acquired characteristics was enforced by the fraudulent claims of the Soviet agronomist Trofim Lysenko, which resulted, among other things, in agricultural mismanagement and the consequent death of hundreds of thousands of people from starvation.

With an ethnically more homogeneous population than the USA, the UK has not had a policy of forced sterilization, although a bishop of Birmingham called for such measures. Nevertheless, many people who were deemed mentally impaired were targeted, including single mothers.[11] The Mental Deficiency Act from 1913, in force until 1959, led to the detention of many thousands, and at its peak, 65,000 were institutionalized and prevented from having children. Although eugenicists were generally traditionalist and conservative, social progressives also supported the efforts. It was in this context that in 1943, C. S. Lewis critically commented that 'what we call Man's power over Nature turns out to be a power exercised by some men over other men with Nature as its instrument.'[12]

In Sweden, Denmark and Norway, eugenics complemented and even contrasted their other social policies. These Social Democratic welfare states supported low earners, yet they also legalized eugenic sterilization efforts from the 1930s to the 1960s or 70s. The term 'racial hygiene', prominently used by the Nazis, has a significant Scandinavian pedigree. People deemed genetically ill, less intelligent, or prone to crime and vagrancy were sterilized. One important aspect, which is relevant to our discussions, is that in the Scandinavian context, these sterilizations required the consent of those being sterilized, though this was often not given in a strictly voluntary, well-informed way. These policies also continued after the war, and from the 1930s to the 1950s, more than 35,500 Scandinavians have been sterilized, though the number may even be much higher. Once again, the majority of victims were women. Historians suggest that Scandinavia's eugenicist past is consistent with the current popularity of pre-natal diagnostic procedures in Sweden and Denmark.

The popularity of eugenics was never championed in a properly scientific way, and inconsistencies in the ideology raise questions about why it was so popular for several decades. In evolution, better-adapted organisms are necessarily more likely to survive and reproduce, while eugenicists suggested that the less intelligent and the physically weaker were likely to have more children. Perhaps the less intelligent (for example) might have more children for other reasons. Yet the eugenics movement suffered from another fundamental scientific flaw. In the 1920s, the consensus began to emerge among geneticists that personality traits are not due to single genes but involve the interaction of several sets of genes.[13] By contrast, eugenicists held on to the idea that biological traits were more simply explained, and they considered that biological markers would be inherited as a unit, producing a trait consistent between people. Davenport studied family trees to find not only a genetic basis for Huntington's disease, which we know to be correct, but also of 'feeble-mindedness', or some combination of learning disability and loose morals. He even hunted for a gene causing a love of the sea, which he thought might run in the families of naval officers!

It is important to note that advances in the science of genetics have done much to discredit many of the principles that underpinned these original eugenic theories. When seen against its flawed scientific theories, the popular eugenics movement of the early twentieth century was highly damaging for any professed eugenic ideals. Nonetheless, scientific insights into the complex relationships between genes, the environment and personality traits did not necessarily make the wider science community unreceptive to eugenics. In a theoretical example, the geneticist Hermann J. Muller, a contemporary of Davenport's, illustrated the inherent problems with applying eugenics to human reproduction and fitness. He supposed that a genetic predisposition for criminality or low intelligence would be inherited recessively. To have the trait in question, a person needs to inherit not just one but two copies of the gene, one from each parent. Many people would have only one copy of the putative gene, which they could pass on to their offspring unnoticed. However, occasionally someone would inherit two copies. Without any genome sequencing technology available at the time, there was no way to identify the 'carriers', who possess only one copy of the gene, which they might then pass on. Muller conceded that in order to reduce the number of people with the undesired, putative recessive trait from 300,000 to 75,000, everybody with that trait would have to remain childless for twenty generations, or for roughly 600 years![14]

In the absence of compelling science, the former practice of eugenics seems like a conspiracy. Why did it gain such prominence? In many of the societies in which eugenics seemed acceptable, the more elite classes were confronted with new social and political developments. The victims of eugenics were notably women and ethnic minorities, which put women of colour in double jeopardy. People with disabilities were also targeted, as too were those who were on the fringes of legality or who resorted to criminality. In contrast, the rich and middle classes dominated the eugenics societies that promoted their repressive policies. Eugenics partly served as a means of maintaining the established order as traditional political arrangements were threatened or became unstable: women gained the right to vote; in the USA, African Americans and Hispanics were kept from political participation; immigrants arrived in significant numbers from Ireland and Italy; and the larger geopolitical situation of the early twentieth century was beset with uncertainty. Eugenics as practised outside Nazi Germany focused on the behaviour of individual persons, especially when this appeared to disrupt the established order, rather than facing the new developments within societies and between nations. By 'weeding out' individual 'defectives', eugenics seemed to relieve society of the moral obligation to improve the educational system and the living conditions of the poor, or to offer opportunities for ethnic minorities to share power.

What does pre-Nazi eugenics have to do with genome editing?

There is a lot of important history here that deserves to be more widely known. Yet of course today's proponents of genome editing, who are predominantly white men, are not proposing anyone's sterilization; and as we will discuss below, it should be clear that advocates of genome enhancements find Nazi politics abhorrent. However, reflecting on the historical, social and political situations may help to identify some of the questions for today's potential uses of genome editing, especially concerning the relationship between biological ideals and socio-political contexts. Although today's societal contexts are not the same, far-reaching uses of genome editing might become as 'useful' in today's socio-political context as eugenics was 'useful' in the early twentieth century. Then, eugenics seemed to provide a way to keep the newly emerging forces in check, in order to preserve the existing political and social arrangements. Nonetheless, individual protagonists who were active in the eugenics societies may have acted from a mix of motivations, some of which may have been progressive and benevolent. However, we should avoid the mistaken suggestion that this historical form of eugenics was primarily or generally concerned with producing genuine medical and social improvements.

In the early twentieth century, the reigning ideology in the West was rather different from the present liberal individualism. The state was often seen as the promoter of civilization, which was notably white and patriarchal. Women were the most prominent victims of eugenic policies, though skin colour and disability were other aspects of eugenic violence.

It is self-evident that the earlier eugenic practices were cruel and unjust. Looking at these Western forms of eugenics, it is not surprising that today's advocates of genetic enhancements reject any overt pressure to 'enhance' the capabilities of children genetically. The troublesome history of eugenics should not be compared with parents' efforts to improve the health and education of their children, and some applications of genome editing may appear to be worthy aspirations. Nevertheless, this earlier practise of eugenics also sought to preserve the status quo in the face of social and political challenges. It is an example of the temptation to tackle social tensions by manipulating other people's minds and bodies. Across the Western world people sought to address the challenges of social and political change with the simplistic ideas of making people's minds and bodies conform to the perceived norms. A question for us today is whether suggestions for genetic enhancements may be driven by worries about social change that could be better addressed by social adaptations rather than physical modifications. If today's proposed genetic enhancements are used to serve a similar social function to the one eugenics did back then – even if today's methods and moral sensibilities are significantly different – then it is likely that they will reflect the features, pressures and expectations of today's society, even though this will not involve any state intervention. A lack of overt state coercion is of course a good thing, but that should not prevent us from seeing the worrying possibility that in subtle ways, people's bodies and minds may be shaped to satisfy the values that are promoted or idealized by wider society. Moreover, the history of eugenics shows that social contexts and goals should not be overlooked when declaring that any new technologies are scientific and objective. There is no doubt that science makes truth claims that can be tested experimentally and must be supported by robust theory. Nevertheless, science and technology are not inherently neutral and impartial, but require critical public reflection. Society's use of scientific findings and technological advances will be influenced by our convictions and interests, in the same way that the selection of particular research questions is influenced by practical goals.

In what sense might today's discussions about biological optimization be influenced by our own socio-economic situation? To what extent might we shape our biology to conform to socio-political trends? In contrast to the early twentieth century, liberal individualism and consumerism, together with competitive capitalism, are the foundations of Western society. Typical proposals for today's genetic enhancements from the philosophical literature include a more powerful memory, a reduced need for sleep, longer lives in better health, athleticism and cosmetic improvements such as greater height and a reduced likelihood of obesity, or all-around competitiveness. Individualism, greater productivity and consumerism are clearly some of the forces behind many of these suggestions. They focus on an individualist mindset within a capitalist economy that already places people with disabilities at the margins. The societal benefits of eugenics will also be influenced by economic arguments. If the lifetime cost of supporting someone with a disability is higher than the cost of genome editing, then there appears to be an economic argument in favour of genome editing. Although we must be realistic about the costs of long-term care for those with disabilities, we should also be uncomfortable about making ethical decisions according to the economic arguments about the worth of people with different disabilities.

While the social pressures have changed since the early twentieth century, there are other ways in which we might shape people's biology to make it compatible with established expectations. This seems to be the new social project for which people might now be tempted to streamline themselves and their children. Even so, while there appears to be a significant connection between the historic function of eugenics and some genome-editing projects, that does not mean that all far-reaching genetic interventions can simply be equated with historical eugenics.

The famous scientist James Watson suggested that parents should be able to choose freely between genetic enhancements. However, personal choice is often not simply 'free' but is influenced in subtle ways by social conventions. We may imagine that we don't need to worry about authoritarian eugenics if the state does not interfere in our biotechnological choices. However, by means of technology, the media, economic and legal procedures, people exercise power over each other in more subtle ways than a government would. For example, skin-lightening products are big business in East Asia. Products for straightening one's hair have been marketed specifically to African American women for over a century. While there is a certain freedom of choice in the use of these technologies, social and cultural pressures with racial overtones add to the mix, and the use of these products can put pressure on those who have chosen not to use them.[15] Society is much more complex than a simple relationship between the state and its citizens, and the government does not necessarily need to be actively involved for power to be exercised. In order to defend freedom and dignity today, we need to reflect on other ways people shape their lives and those of others in order to conform to society's unspoken values.

Nazi eugenics

It is clear that today's efforts in human genetic modification cannot be equated with historical eugenics, especially if we take proper account of Nazi eugenics programmes. In the first place, Nazi eugenics included the systematic murder of millions of Jews and of other minorities, propelled by the ideology of 'racial hygiene'. Today's advocates of genetic enhancements are clear in their criticism of this abhorrent, violent and racist ideology. Those advocating various uses of genome editing are of course not suggesting compulsory sterilizations, which were often practised under eugenics banners, even outside of Germany.

Written in 1925, Hitler's manifesto *Mein Kampf* stated that 'The stronger must dominate and not mate with the weaker.' Some German eugenicists had called for the killing of people with severe mental disability as early as 1920, and after assuming power, the Nazis created a legal basis for the forced sterilization of people with hereditary disease in the 'Law for the Prevention of Hereditary Defective Offspring' (1933). In part the law provided overtly racist reasons for legalizing sterilization and abortion in order to 'tend the Aryan gene pool', and the Nazis established genetic health courts in which two doctors and a lawyer determined who should be sterilized in each case. In 1935, the Nuremberg race laws prohibited marriages between 'Aryans' and Jews. However, racism was not the only driving principle, and by 1939, the Nazis had launched a secret operation to kill disabled newborns and children under the age of three, which was later extended to include children and adults with disabilities. People with disabilities were often described as 'lives unworthy of life'. In this way the Nazis murdered about 200,000 people with disabilities. These state-sanctioned murders were overseen by medical experts, professionals and social activists. Sometimes the practitioners were even social progressives. For example, even before the Nazis rose to power, some who started calling for people with disabilities to be killed also opposed the death penalty and supported greater rights for women. Many also saw their actions as arising from science.

Here Nazi eugenics confounds some of our common assumptions. The killings of people with disabilities that began in the late 1930s were not motivated by racist ideology or planning for the future biological welfare of the nation. A memorandum Hitler received from his personal physician shows that the important point was to save the money that would otherwise be spent on the care for the disabled and the mentally ill. The memorandum also suggested that the general public feels revulsion when seeing people with severe disabilities. The draft of a law allowing for the murder of disabled people was more widely circulated in 1940 and came into effect in 1941.[16] However, by that time the elaborate bureaucratic and institutional apparatus had already killed about 70,000 people with intellectual disabilities and mental illness – Germans who would otherwise have been counted as 'Aryans'. The Nazis introduced the term *euthanasia* for this kind of murder (Greek for 'good death'). In German, *Euthanasie* still means the murder of a person with disability as practised by the Nazis, without any connotation of a 'good death'.

Importantly, expressions of racial ideology are absent from documents legalizing the mass murder of those with disabilities. Neither did internal communications within the medical and bureaucratic apparatus use racial terms. The official questionnaires intended to identify victims asked for many things, but not for their 'racial' background or even for factors that might indicate a heritable nature of their condition. The disabilities in question sometimes had a genetic cause, but often they were due to complications in giving birth or to accidents and war injuries. Ethicists debating genome editing sometimes talk about the 'gene pool' that the Nazis supposedly wanted to protect, and suggest that this was based on racial ideology. However, the facts reveal that considerations of race and heritability were not relevant as far as Nazi 'euthanasia' of those with disabilities was concerned. In preparing to murder people with disabilities, leaders of institutions did not ask questions about race or heredity. Rather, they wanted to know whether the ill and the disabled contributed to the economic operation of their wards and if any economic productivity was likely. This mass murder of those with disabilities was primarily ableist, not racist.

Perhaps the most worrying aspect about the Nazi murder of people with disabilities is that this practice was fairly widely tolerated. Many parents were undoubtedly shocked that their disabled or mentally ill relatives were now dead, and care homes often deceived parents by inventing natural causes of death. Nevertheless, many parents must have seen through the deception but pretended that nothing illicit had happened, while other families had already cut off contact with their disabled relatives. Institutions were allowed to kill their wards, not legally at the beginning, but by the Führer's personal permission, though by no means were they obliged to do so. Often a relatively small protest by relatives was enough to prevent the murder, and parents who inquired about the whereabouts of their disabled children were allowed to save them from the gas chambers by taking them home to live with them. The standard routine in transporting people with disabilities to the gas chambers often included deliberate delays to give relatives a chance to intervene. Such a practice would have been unthinkable for the Nazi murder of the Jews. Nevertheless, many families and ordinary citizens were complicit in the mass murder of disabled people, and some parents even asked their local psychiatrist for a 'mercy death' for their child. The only German institution that consistently condemned this mass murder was the Roman Catholic Church, and the protestations of one prominent bishop brought the mass murder to a temporary halt. 'Where protest did erupt', a historian notes, 'it hardly ever stemmed from the principles of modern political checks and balances or the ideas of a secular humanism, but from a faith, long diminished, in the creation of every human person in the image of God – no matter how crippled, idiotic or feeble-minded, reliant on care or suffering.'[17]

When discussing genome editing and eugenics, several philosophers have commented on the Nazi murder of people with disabilities. They blame this on the Nazi race ideology that sought to 'improve the gene pool'. They then suggest that something like that would be unlikely to happen today, for three reasons. First, for the large majority of people, race would play no role in the deployment of genetic technologies, whether medical or enhancing. Second, people today do not see themselves on a wider mission to reduce the incidence of genetic conditions in the future generations of the wider population. And third, we are not Nazis. They therefore feel confident that Nazi eugenics is not particularly relevant when discussing present genetic technologies. However, these arguments are flawed. The Nazi murder of 200,000 disabled people was not due to questions of race. Neither was their murder motivated by considerations about future generations that might inherit a genetic impairment. An ableist reaction against people with disabilities, along with economic reasons, played an important role. Finally, as we have noted, this was not merely sanctioned by Nazi ideologues but a significant number of ordinary citizens were also complicit in the process, and many tolerated or actively supported the murder of their disabled relatives.

Of course, there are obvious differences between promoting genetic germline editing and killing or sterilizing people with disabilities. We are certainly not suggesting that genome editing or enhancements would result in such abhorrent practices. It is clearly inappropriate to equate genetic enhancements, or even genetic modifications on the whole, with this form of eugenics. We assume that people with disabilities will not face overt and systematic violence but will continue to benefit from robust legal protection. Nevertheless, the example of Nazi eugenics suggests that there is a deep-seated prejudice against people with disabilities, which can sometimes erupt in violence, physical, verbal or otherwise. So much is clear even without the evidence from Nazi eugenics, and sometimes even ordinary citizens look past such injustices.

Several commentators have suggested that parents should be able to choose genetic enhancements for their offspring, voluntarily and without interference from the state or any other authorities. However, such an unencumbered libertarian and individualistic approach avoids asking important questions. Historical information about eugenics raises serious questions of whether far-reaching genome modifications, especially genetic enhancements, will reinforce an ableist mentality, which assumes that independence and physical functioning should be maximized, and that dependence, weakness and vulnerability are defects that are inherently bad. People's natural inhibitions usually keep them from expressing degrading views or carrying out harmful actions towards people with disabilities. However, these inhibitions sometimes fail, given the wrong circumstances. There is no reason to expect atrocities on the scale of the Nazi mass murder of people with disabilities, but this should warn us of the danger of taking anti-disability prejudice lightly.

Conclusion

Eugenics has taken on different manifestations at different times and places, but our historical survey has shown that it leaves nothing that we would want to emulate. The idea of an enlightened, 'liberal' or 'consumer eugenics' is misleading, not only because eugenic policies have been consistently repressive but also because people with liberal views have often been complicit in authoritarian violence against women, people of colour and people with disabilities. Because of this violent and coercive history, the use of genome editing for regular medical purposes should not be equated with eugenics. In contrast to twentieth-century eugenics, which has been described as a 'war on the weak', it has been suggested that genome editing in the twenty-first century should be described as a 'war for the weak'. Nonetheless, a disproportionate pursuit of perfect health still raises serious questions. While a pregnant woman will of course act to ensure the best for her child, we should not conclude that all ways for pursuing health and strength, such as those involving genome editing, are legitimate. During the twentieth century the elite classes used eugenics to preserve the status quo, rather than facing uncomfortable challenges to existing power structures by women, people of colour and people with disabilities. Similarly, today's ideas of 'consumer' or 'liberal' eugenics will probably ignore or reinforce some of our current socio-economic problems. Today's individualist mindset is of course very different from that of the eugenicists in the early twentieth century, who were white, openly patriarchal and chauvinistic. Yet then as now the human body has been an area for promoting conformity to social expectations. Today's suggestions that we should permit personal biological enhancements tend to idealize the principles of individualism and productivity. These already place many people with disabilities at the margins of society, and such a goal of physical perfection will only serve to exaggerate existing differences. The view that every human being has special human dignity, regardless of differences in their physical functioning or intelligence, contrasts with some of the manipulative social strategies for making 'better people'.

Present-day advocates of genetic enhancements argue that their desired improvements would be very different from Nazi practices. However, they miss the point if they mischaracterize the Nazi eugenics ideology. The Nazi mass murder of people with disabilities was not motivated by *race* and it was not concerned with improving *future* generations (the 'gene pool'). Instead, the main motivation seems to have been visceral prejudice against disabled people, combined with economic considerations of their dependency on others. To a worrying extent their abhorrent policies were supported by *ordinary* citizens, physicians and nurses. An increase in prejudice, even in subtle ways, may therefore not be quite as far-fetched as we might like to think, even today. We are not suggesting that genetic modifications and enhancements will lead to murder or any other form of physical violence against people with disabilities. Yet it is possible that the new striving to enhance physical functioning could well reinforce existing prejudices against those who are less able.

A recent example of such prejudice against people who differ from society's norms is seen in the case of a company selling T-shirts with the slogan 'Let's Make Down Syndrome Extinct'. This received public outrage, especially from the parents of children with Down's. One mother, the actress Sally Phillips, commented: 'Eugenic ideas are really taking hold – the idea that there is this subclass of humans and it is better we get rid of them.' Although surveys show a high quality of life among people with Down's, the historian and journalist Tim Stanley argues that society is attracted to 'making life as perfect as possible', with biotechnologies promising 'no birth defects at all'. Eugenic strategies may seem like a relatively easy option to eliminate this condition, but 'the easiest course doesn't necessarily make for a happy or fulfilling life … The true moral test of a society is not how pretty, sober or well organised it is – but how it treats its most vulnerable, even its most difficult, citizens.'[18]

There may be far-reaching genetic interventions that should not simply be equated with eugenics, which was usually coercive and which targeted specific groups, especially women, people of colour and those with physical or mental disabilities. However, when debating genome editing, concerns relating more broadly to historical eugenics cannot be simply dismissed as alarmist. On the one hand, using the term 'eugenics' as a simple rhetorical device for vilifying people who promote genome editing and enhancements is inappropriate and fails to do justice to history. However, there are lessons to be learned. Genetic enhancements might be used to promote social norms for streamlining the human body to conform to current prejudices, pushing those who are less self-reliant, conventionally unattractive or less productive to the margins of society. It seems likely that a society that chooses to spend more time and money on screening or editing will be less interested in caring for people who do not join in with these programmes. Of course, efforts to prevent or mitigate severe diseases should not be rejected out of hand, yet we should be careful to ensure that these do not result in policies that stereotypically maximize health and functioning, while ignoring significant social contexts. Writing more than a decade before the discovery of genome-editing technologies, David King warned of laissez-faire eugenics in which this type of control has crept in unnoticed or unintended: 'The danger we will need to guard against is the development of a kind of eugenic common sense, that it is irresponsible to refuse to undergo tests, and that every child has the "right" to a healthy genetic endowment.' An emphasis on biological means for promoting a successful, healthy and fulfilling life may tempt some people to claim privilege as their natural right and to feel a diminished responsibility for improving the educational, social and political opportunities for those on the margins.

With the discovery of new ways to modify humans, by transforming their genes rather than through selective breeding or murder, it is not overly alarmist to say that eugenics could return, but in a new private form, shaped by the dynamics of a democratic consumer culture. The new eugenics is more likely to be driven by individual preferences in which parents will choose to edit their child's genes to prevent illness or to improve their strength or looks. In this new scenario, there is a danger that the encouragement to make changes may not come from government directives but could be promoted by the biotech industry that developed the technology. We must also guard against any pressure towards editing human embryos, for, as the author Adam Cohen has noted, even in the absence of overt state policies, it is not such a great a leap from 'you *can* have a genetically improved baby' to suggestions that it would be irresponsible to have a disabled child and therefore that 'you *must* have a genetically improved baby'. 'Renegade scientists and totalitarian loonies are not the folks most likely to abuse genetic engineering', the ethicist Arthur Caplan commented in *Time* magazine: 'You and I are, not because we are bad but because we want to do good ... parents understandably want to give their kids every advantage. ... The most likely way for eugenics to enter into our lives is through the front door as nervous parents – awash in advertising, marketing and hype – struggle to ensure that their little bundle of joy is not left behind.'

Notes

1 Kevin Smith, 'Time to start Intervening in the Human Germline? A Utilitarian Perspective', *Bioethics* 34 (2019): 90–104.

2 Nicholas Agar, *Liberal Eugenics: In Defence of Human Enhancement* (Malden, MA: Blackwell, 2005).

3 See Kurt Bayertz, *GenEthics: Technological Intervention in Human Reproduction as a Philosophical Problem*, trans. S. L. Kirkby (Cambridge: Cambridge University Press, 1994), ch. 2.

4 Almut Caspary, 'The Patristic Era: Early Christian Attitudes toward the Disfigured Outcast', in *Disability in the Christian Tradition: A Reader*, Brian Brock and John Swinton, eds (Grand Rapids, MI: Eerdmans, 2012), p. 28.

5 Edward J. Larsen, 'Biology and the Emergence of the Anglo-American Eugenics Movement', in *Biology and Ideology: From Descartes to Dawkins*, D. R. Alexander and R. L. Numbers, eds (Chicago and London: University of Chicago Press, 2010), pp. 165–91.

6 Steven Shapin, 'The Virtue of Scientific Thinking', 20 Jan. 2015, *Boston Review*, http://bostonreview.net/steven-shapin-scientism-virtue (accessed 22.9.2020).

7 T. H. Huxley, 'Evolution and Ethics', pt. 2, *Popular Science Monthly* 44 (1893), 178–91.

8 On eugenics in particular countries, see the *Oxford Handbook of the History of Eugenics*, A. Bashford and P. Levine, eds (Oxford: Oxford University Press, 2010).

9 Charlotte Hunt-Grubbe, 'The Elementary DNA of Dr Watson', *The Sunday Times*, 14 Oct. 2007.

10 Charles Davenport, *Heredity in Relation to Eugenics* (New York: Holt, 1911).

11 See the BBC documentary 'Eugenics: Science's Greatest Scandal', with Angela Saini and Adam Pearson (2019).

12 C. S. Lewis, *The Abolition of Man: Reflections on Education with Special Reference to the Teaching of English in the Upper Forms of Schools* (London: Oxford University Press, 1943), ch. 3.

13 Garland E. Allen, 'Is a New Eugenics Afoot?' *Science* 294, issue 5540 (2001): 59–61.

14 Hermann J. Muller, *Out of the Night: A Biologist's View of the Future*, 1936 (according to Bayertz, *Genethics*, ch. 3).

15 Carl Elliott, *Better than Well: American Medicine Meets the American Dream* (New York: Norton, 2003).

16 For the Nazi mass murder of people with disabilities, see the book by the German historian Götz Aly, *Die Belasteten: 'Euthanasie' 1939–1945. Eine Gesellschaftsgeschichte* (Frankfurt/Main: Fischer, 2014).

17 Aly, *Die Belasteten*, p. 22 (trans. A. Massmann).

18 Tim Stanley, 'Down's Syndrome People risk 'Extinction' at the Hands of Science, Fear and Ignorance', *The Telegraph*, 18 Jan. 2016, https://tinyurl.com/y3w2dex4 (accessed 19.1. 2021).

6

Being human in an age of biotechnology

Our discussions in the previous chapters suggest that there are very few ethical issues associated with genome editing of somatic cells – modifications done in children or adults, which would not be passed on to future generations – for *therapeutic* reasons. The consensus among scientific experts is that the risk of unintended genetic changes is presently the clearest reason not to modify human embryos, even if it is done for serious genetic conditions. These technical obstacles may be overcome at some point in the future, but it is difficult to assess whether a procedure is safe enough to proceed, given that the stakes are very high in embryo modification. They will be high especially when there is an established alternative to germline editing, such as pre-implantation genetic diagnosis (PGD). A report in the journal *Science* cites a geneticist making the same observation: 'I continue to struggle to imagine plausible situations in which clinical germline editing provides a path forward to address an unmet medical need.'[1]

There could be some very rare cases in which PGD is not a viable alternative, yet even then the question is whether maximizing therapeutic potential via germline editing is the path that we should follow. Of course, medical treatment must be a high moral goal, but this can involve a trade-off. A push even for small medical gains could further entrench the view that chronic illnesses or disabilities that remain beyond a cure are fundamentally undesirable. The point is not that individual A might take personal offense at individual B for pursuing medical treatment. Rather, there is a wider societal dimension to human life. The media report new treatments as progress, and together with researchers and pharmaceutical companies they reinforce the narrative of the forward march of science. Meanwhile, we talk less about the accessibility of buildings and buses or the inclusion of people with disabilities in regular social activities. Germline editing will do nothing to improve this, and might indeed reduce the readiness to accommodate people who will continue to have different needs and desires. Beyond the medical level, it is possible that germline editing will open the door to enhancements that promise to increase physical abilities or alter our mental and emotional states – a procedure that is fraught with even greater moral issues.

Starting with the basics, having a human genome is of course a necessary part of being human, but it alone is not sufficient to define us. In and of itself, DNA is just another chemical polymer, and its sequence is only endowed with any meaning by the context in which it is located. When left on the laboratory bench, it will do nothing but slowly degrade. It has a very different role within a living cell, in which it is a template for making all the body's constituents. However, a cell by itself is still not sufficient to be defined as human, and it only has significance in the context of a particular tissue or organ, which in turn will only function in its context within the whole organism. We are not single-cell organisms! The emerging area of epigenetics is also showing that our DNA is not a simple blueprint, but that many external factors determine which genes are switched on or off. At a higher level still, the individual human assumes a different significance in the context of a community. The ways in which our brains develop are not specified by our genes but depend on repeated interactions and learning. We would not be able to speak or to walk on two legs without encouragement from a young age, and our ability to learn continues throughout life – that's not simply in our DNA. We are communal and social beings and we cannot properly understand ourselves without referring to any of these contexts.

As new genetic techniques for changing our minds and bodies become available, we need to ask what types of physical bodies and mental capacities we desire, what sort of people we want to be and what kind of society we want to live in. In part the questions are for each of us as individuals. Yet nobody decides for themselves in isolation, as our choices affect our children and future generations, as well as wider society. All of this raises the fundamental issue of what it really means to be human. This subject has occupied philosophers for millennia, and the answer is likely to become less clear as genetic tools allow us to modify our bodies and our minds. Some people suggest that greater physical and mental capabilities lead to better lives, while others are quite content with their condition and live fulfilled lives in spite of impairments. Because our bodies play an important part in defining our identity, alterations to their structure or function may also affect how we perceive ourselves. In this final chapter, we will take a broader look at what it means to be human in the context of these questions. Of course, we should remember that genetic manipulations are not the only way in which our identities can be altered. A traumatic brain injury can produce dramatic changes in behaviour, while sport and exercise have their own effects, and even the process of learning causes a rewiring of regions in the brain. However, genetic technologies provide new and irreversible ways for deliberately remoulding our identities.

How, then, does changing a DNA sequence affect our humanness? We are more than the sum of our genes; we are also shaped by our environment, our experiences, the people with whom we interact and even the things that we eat. Our physical bodies as well as our social communities are very important for defining who we are. Along with the rest of creation they are described as very good (Gen. 1.31). We are not just 'spiritual' beings, for which fulfilment is achieved by escaping from our bodies. It would be misleading to imagine that our bodies are merely the means through which our identities relate to the 'external' world.

As with other animals, our bodies are part of God's good intentions for us human beings. However, we often misunderstand what it means to be embodied and ignore the limitations that this necessarily involves. For example, proponents of human enhancement often assume that it would be good to increase mental traits that are considered to be desirable. Our memory, which some wish to 'improve' with genetic enhancements, is a good example, and there may indeed be some situations where a more powerful memory would be a good thing. However, that is often not correct and it fails to understand how we live meaningful lives within our natural limitations. Despite what might be claimed, the normal process of forgetting is not primarily a deficit, but an important feature that enables us to separate what is significant and valued from what is trivial. We consolidate our memories, and unimportant things are forgotten. This is a selection process that is import-ant for the meaning and significance of our experiences, but it occurs largely without our conscious control. By contrast, if we were to be enabled to remember more, due to some enhancement, this would alter our standards of what is worth remembering and what matters in life, and it is not certain that our standards would change for the better! We might end up remembering things we would prefer to forget. The suggestion that to remember more is better is inspired by the idea that our memories work like computer hard drives and that greater capacity will always be better. However, a hard drive does not select which are the important things for us to retain, as our brains do. The computer is just a passive machine, and it is the user who selects files for the hard drive to store. If I were fundamentally like a computer user, controlling a hard drive to control my memory, then I would be like a spirit, separate from my body. I would be like a 'ghost in the machine'.

The biblical Scriptures value our finite bodies so much that they even call the body a 'temple of God'. Here the body is not only appreciated by its remarkable, albeit mundane functions, but an even more dramatic claim is made: our humble body is precisely the entity that engages with eternal, divine things, expressing God's worship and praise. Indeed, Christians see the ultimate affirmation of the human physical form in the incarnation, in which God himself took on human flesh.

Human dignity

The significance of our bodies raises the important issue of human dignity and how it is affected by any biotechnological interventions. The idea that dignity is non-negotiable and shared by the entire human species is included in the 1948 Universal Declaration of Human Rights, which states that 'All human beings are born free and equal in dignity and rights.' Since we are embodied beings, we also need to think about the dignity of the human body and to ask how biotechnological interventions affect human dignity.

The life sciences describe the human person in mechanistic terms. Of course, genes and brain circuits make a significant contribution to what and who we are, but 'human dignity' also makes further moral claims that we cannot test experimentally, and we need to resist the temptation to reduce a human person entirely to a description in mechanistic terms.

Human beings may not behave quite as rationally and freely as philosophers have assumed when making arguments about human dignity. Nevertheless, we have unique features that indicate our special status within nature. Basic moral respect for others is a critical aspect of this. Animals like chimpanzees show empathy and practise cooperation, but non-human animals do not display the same moral regard for other members of their species that humans intuitively do (or *fail* to do). That isn't a moral failure in chimpanzees, but emphasizes that there are both qualitative and quantitative differences between us and other living things, despite our genomic similarities.

While most people agree that dignity is important, it is not easy to define, and there are contrasting ideas of what it means and what it implies for any biotechnological interventions. We can think of many instances in which people have been treated badly, yet they have retained their dignity, even in dehumanizing conditions. Dignity also depends on the context in which it is considered. In medical practice, we are often required to undergo procedures that might be described as 'undignified' in different social contexts, yet we know that this is ultimately for our own benefit. Nobody can lose their human dignity, even if they appear stupid or evil. On the other hand, the bioethicist Ruth Macklin wrote an editorial in 2003 entitled 'Dignity is a Useless Concept'.[2] For Macklin, who has stood up for the humane treatment of people, talk of dignity achieves nothing beyond affirming personal autonomy; it is a restatement of the fact that nobody has the right to impinge on the life, body or freedom of another. Human dignity requires us to appreciate the way others make choices for themselves for personal reasons; it demands nothing more and nothing less. We respect others, Macklin submits, for the personal decisions that they freely make, based on good information, as long as they do not harm others. By contrast, we maintain that the concept of dignity is an important guiding principle that cannot be reduced to respect for autonomy.

In our affirmation of dignity we agree with Macklin that it includes our moral obligation to recognize people as much more than objects, but as persons, subjects with wishes, personalities and emotions. Infringements of human dignity are then characterized by treating subjects predominantly as objects – examples of which include disenfranchisement, deprivation, humiliation and even torture.

The Enlightenment spelled out the meaning of human dignity in terms of autonomy, though it understood the idea in varying ways. The word 'autonomy' means that people are not simply subordinate to the laws that are imposed on them, but they 'themselves' (Greek *auto-*) become makers of the 'law' (*nomos*). In the West, one of the major cultural changes in the last centuries has been a trend towards greater personal autonomy and self-determination, which is seen in many things like the choice of profession, religious freedom and the right to vote. Christians have good reasons to support this quest for personal freedom in the light of biblical themes such as the liberation of Israel from slavery or Paul's emphasis on freedom in Christ. Personal autonomy also implies that genetic modifications – or even research on someone's genome – must not be done unless the affected person makes a free, well-informed decision.

Nevertheless, we consider that human dignity involves other aspects that are not captured by freedom of choice alone. First, while respect for autonomy prohibits us from encroaching on other people's freedom, human dignity also includes a moral call to work for the *benefit* of others. This is a moral and not a legal requirement. Second, such a view of human dignity obliges each of us to respect our own dignity, and to take care of our own bodies and look after our health. In certain circumstances it is also legitimate to refuse medical treatment, but not all 'autonomous' choice is morally appropriate, even if it does no harm to others. For this reason, society cannot allow biotechnological procedures, such as genetic enhancements, to be marketed freely if there is a significant health risk, even if the choices that individuals would make about them are free and well informed. Human dignity not only protects one person from another's encroachment but also constrains our own personal choices. Those who *reduce* human dignity to the ability to make informed decisions suggest that free choice itself is the critical moral feature of being human. However, as we have discussed, human dignity consists not only in decision-making but includes the dignity of the living human body, even apart from decision-making. If, by contrast, dignity is reduced to nothing more than informed decisions, what about the dignity of people whose ability to choose is restricted, perhaps due to dementia or severe disability?

Is the human genome sacred?

The human genome alone does not define what it means to be human, but it plays an important role in a biological description of our species. The genome has also been linked to questions of dignity. It has been argued that since the possession of a human genome is an essential characteristic of all humans, to modify it would be an affront to natural human dignity. This line of thinking is associated with UNESCO's 1997 Universal Declaration on the Human Genome and Human Rights, which states: 'The human genome underlies the fundamental unity of all members of the human family, as well as the recognition of their inherent dignity and diversity. In a symbolic sense, it is the heritage of humanity.' The Declaration does not use words like 'sacredness' or 'sanctity', but the argument could be made that the genome should be preserved in its 'natural' form, just as the historic character of a UNESCO world heritage site, for example India's Taj Mahal, should be kept intact and must not be sacrificed for the sake of some other benefit, financial or otherwise.

We challenge this assumption on several grounds. In the first instance, talk of preserving humanity's genetic heritage assumes that there is a single static entity such as *the* human genome. However, everyone's genome is subtly different, varying at about one position in every thousand (i.e. with about 3 million differences in the three-billion-letter genome between people). Genetic variation is simply another example of our individual human uniqueness, and by nature, new genetic differences arise randomly in every generation. A child inherits a new, random combination of the parents' genes, and in addition, genes are subject to natural mutations at a rate of about 30 random changes per generation.

Genetic diversity is a fact of nature, without which evolution would have ground to a halt long ago. Some people may argue that altering genes should be left exclusively to God or 'nature', and that these natural processes are more reliable than any attempts at human editing. This too does not withstand scrutiny. If we have the means to prevent a genetic disease that would come about naturally, then surely this can be a good thing – if indeed it is less risky for the future of the human species than random natural mutations.

Moreover, *therapeutic* genome editing would not introduce any new DNA sequence information into the human genome, but merely return a mutant form to what is found in the rest of the population. Many of the proposed *enhancements* would not produce new gene variants either, but merely modify the natural distribution of existing variations: an edited individual would be provided with the version of a gene that is naturally present in other people, with the desired attribute. This would not produce any new genes, though in *one* person it might generate a new combination of genes that occurs naturally in *other* people – regardless of whether this is done in embryos or adults. Editing an embryo's genome to restore the healthy version of (say) the gene that, when mutated, causes cystic fibrosis, does not introduce any new variants into the pool of human genes. One could perhaps argue that genetic modifications would reduce the diversity of the human genome slightly if a certain disease mutation is corrected in a number of people. Yet here the fact of genetic diversity is morally irrelevant – the question is not what is better for the statistics, but what is better for people. For example, people of African or Caribbean descent suffering from sickle-cell disease – a condition more common in those regions – may have a genuine interest in a therapy that modifies their genes, and here it would be pointless to lament that that would make them less African or Caribbean. This approach is indeed reflected in UNESCO's 2005 Universal Declaration on Bioethics and Human Rights, which states: 'The interests and welfare of the individual should have priority over the sole interest of science or society.'

The image of God: human beings as God's vice-regents

The question of what it means to be human has received one classic answer in the biblical tradition that sees humans as created in the image of God. The Bible also speaks in other ways about the human person, but the image of God stands out as a prominent concept from its very first chapter, and the theme reappears in the New Testament. Although we do not find the expression 'human dignity' in the Bible (it originated in the Greco-Roman world), the concept of the 'image of God' has contributed significantly to the concept of human dignity in Western culture. However, we need to think carefully about what the image of God means. It clearly does not refer to the attributes of our physical bodies, for God is not flesh and blood like humans. The biblical authors did not think of God as walking upright on two legs or as biologically male and female. What, then, is the 'image of God'? The most prominent text containing this expression is Genesis 1.26–27:

> Then God said, 'Let us make humankind in our image, according to our likeness; and let them have dominion over the fish of the sea, and over the birds of the air, and over the cattle, and over all the wild animals of the earth, and over every creeping thing that creeps upon the earth.'
> So God created humankind in his image, in the image of God he created them; male and female he created them.

In previous centuries much Christian writing on the image of God focused on our specific human capacities, such as rational thought and moral responsibility. However, our understanding of this phrase has changed in the light of what such an image meant in the Ancient Near East, where it originally came from. The Hebrew word for 'image' in the passage above refers to a physical statue, standing for someone as their representative. In interpreting the 'image of God', biblical scholars have focused on the way people understood such statues. In the Ancient Near East, statues represented deities in temples, and kings were similarly represented by statues in public spaces. Humanity is then created as the representation of the creator God within wider creation. Humanity as a whole is even given the particular role that was reserved for kings in antiquity: to reflect God's beneficent intentions in governing creation as God's earthly representative (see also Ps. 8). When Genesis describes humanity as created in the image of God, all humans have the same vocation as a king or queen. Regardless of ethnicity, gender or other characteristics, all humanity is God's vice-regent. Yet the special honour also implies a mission: to look after the world and to keep destructive forces at bay. In ancient times, the king was understood to enact the systems of social justice and to work against the unpredictable forces of chaos that hamper human life. In a much later example, the English abolitionist William Wilberforce was inspired by the 'image of God' in his fight against slavery.[3] One contribution of the image of God to the moral concept of human dignity consists in the call to show concern for other creatures, human and non-human, rather than merely to refrain from encroaching on the liberties of other people.

In this practical sense, the creation of humanity in the image of God points back to God as a creator who desires an order in which living beings prosper and thrive, and in which humans play their part in promoting that order. Such 'dominion', as the cited passage calls it, can of course be abused and have harmful effects. Nevertheless, when the text was written about 2,500 years ago, its authors were much more aware that the natural world was a mysterious and dangerous place. The call for humans to tend the world gives us a mandate to work to alleviate pain and suffering. God created a good, functioning biosphere, yet nature is never static, and some parts are chaotic. Our vocation as God's image-bearers is to tame the destructive forces and to restore order where chaos threatens life – with cancer or the tuberculosis bacterium, for example. God is creator, and we are co-creators with him. As far as humanity today is concerned, science is one of the ways in which we fulfil that calling. Even the biblical expression of 'subduing the earth' (Gen. 1.28) makes sense as we work to cure disease and alleviate suffering. The image of God is less a physical or mental attribute than a divine calling.

While the mission of humans as rulers remains an important way of thinking about humanity, we should also guard against anthropocentric arrogance. The human mandate to promote order in creation is not to be exercised in arbitrary human choices but in responsibility to God, as the book of Genesis goes on to narrate (see Gen. 6). In other places, the Bible also points to human limitations and weaknesses in portraying us as 'dust that breathes'. The Apostle Paul sees the 'image' less in 'dominion', but particularly in Jesus Christ, who lived a life of serving others. Human decisions are also to be measured critically against how well they serve others.

By supporting life against destructive chaos and so representing God's activity on earth, humans are to show his loving character. In today's biomedical ethics, this perspective lends force to the medical protection of life through efforts against accident, disease and frailty. Some will ask whether, with biomedical innovation, we are 'playing God'. First, most scientists are not playing; in a significant sense, they're doing scientific work for the good of society and the advancement of knowledge. Second, God is creative, and in some small way we reflect his creativity as we tend his creation.

Moreover, statements about the image of God are contained within passages that have a keen, if pre-scientific interest in the natural order that is expressed in biological terms. Genesis 1 mentions the 'kinds' of living things ten times, and both humans and non-human animals are explicitly 'blessed' with biological reproduction. The text uses ideas like species and reproduction, albeit not in a technical biological sense. Obviously, people with disabilities are the children of human parents, members of the same 'kind' as their parents, so the conclusion must be that they are no less human than anyone else, enjoying human dignity as God's image.

The foundational principle is that every human being is made in God's image, regardless of the way that their genetic endowment might affect their capacities. This gives a clear mandate to respect the dignity of every person. Although we employ scientific discoveries and new technologies to overcome disease and infirmity, we must not do so in a way that dehumanizes either the recipients of any treatment or the scientists and medics who administer it. We should also protect the dignity of bystanders whose value and social standing might be reduced by the use of biotechnological procedures. Although many medical interventions are good and helpful, a zealous commitment to increased levels of function can have the effect of excluding other people.

So a Christian assessment of genetic therapies must begin with a recognition of the humanity of each individual involved. This means that scientists, theologians and ethicists must work together with affected individuals and stakeholders to understand the relevant condition, before assessing if and how genetic science might be able to help. In the course of this interaction, the opinion may also be challenged that some genetic conditions are so severe that they clearly reduce the quality of life.

Although we are made in the image of God, we are not God.

As we explore new technologies, we must be active as God's collaborators, but not misuse that power and not desire to be God. The Christian doctrine of sin has traditionally focused on the dangers of hubris and arrogance. This remains a valid concern, although the doctrine of sin also includes anything that works to counteract faith, hope and love, including aspects of complacency and passivity. As we try to discern the ethical principles of genome editing, we will need to determine whether the things that we do are too passive to reflect God's image or whether, by our overactivity, we are attempting to take God's place. An important part of being made in the image of God is human free will, though God both empowers and sets limits on our freedom. Matters are a lot more complicated than merely affirming that we should exercise a free, informed choice that does not hurt others.

The image of God: human beings defined by personal relationships

The understanding of humanity as God's vice-regent highlights human authority and honour. Yet at the same time, another aspect of humans as created in the image of God is that of our relationships. After each of the works of creation was declared 'good' in the opening chapter of Genesis, the second chapter contains the first example of something that is 'not good': 'It is not good that man should be alone' (Gen. 2.18). Genesis 1.27, after saying that humans are created in God's image, states that humanity is created 'male and female', in relationship. The other animals that God created – fish, birds, cattle – are commanded to reproduce, yet they are not explicitly described as male and female. In contrast, sharing life together, beyond mere subsistence and reproduction, sets humanity apart from animals in this perspective. The warmer, affectionate appreciation of the other person, here seen in the husband–wife relationship, defines a part of what it means to be human. Later mentions of the 'image of God' in Genesis 5 and 9 further extend the range of significant relationships from husband and wife to the family and even to society on the whole. Our relationships remain a fundamental human need regardless of our genetic endowment. Indeed, relationship is implied in God himself as the Trinity, Father, Son and Holy Spirit in perfect harmony.

The philosopher Martha Nussbaum strikes a similar note about relationships when, in a secular context, she reaffirms the ancient view that the human person is fundamentally a 'social animal'.[4] She critiques the Enlightenment philosopher David Hume, according to whom moral obligations only arise between equals who have the capacity to hurt each other, but who have more to gain from cooperation. These are two important perspectives on human relationships. The idea that moral obligations are strictly reduced towards people who are not 'roughly equal' goes too far but, in keeping with Hume, we should not simply denounce the cooler interaction between peers who affirm their own interest and a degree of interdependence. On the other hand, Nussbaum rightly insists on the significant social relationships that are tightly woven into human identity, with an interdependence that also implies vulnerability. These two different perspectives on our humanity mirror the dual perspective on humanity that is evident in Genesis 1: the royal and the relational understanding of the image of God.

The Enlightenment emphasis on autonomy highlights human rational faculties, but we are not just rational creatures; we are also beings with emotions and warm personal ties. This point is important also for the dignity of people with a cognitive impairment. Being human is much more than having autonomous unencumbered choice, as our interdependence and relationships also imply compromise and vulnerability. Close personal relationships will inevitably limit our free, unencumbered choice, but that makes us more human, not less.

An important part of being human is evident in the parent–child relationship. There are very real questions about whether germline manipulation, especially for enhancements, would fundamentally change this dynamic. If an embryo's genome is altered to suit the parents' preferences, the child can easily become more like a product or a project with specific expectations, governed by the parents' choice about what they deem best for the child – rather than being welcomed as an unknown precious gift. In most caring parental situations, children are loved and cared for regardless of their genetic makeup. With genetically enhanced offspring, there is a danger that this could change, as parents extend their ability to design. If parents enhance their children as embryos, there is a risk that they will become more like commodities that have been ordered at will. This is not alarmist speculation, as the detrimental effects that 'helicopter parenting' has on 'overscheduled children' have been documented. The parent–child relationship will also be compromised if the design does not turn out as expected or if the child later questions their parents' choices. I may have inherited physical or mental characteristics from my parents by natural means, but they can hardly be blamed for this. In contrast, the parents of a genetically enhanced child are directly responsible for introducing some of the child's characteristics while excluding others.

While parents inevitably and legitimately make plans and choices for their children, parental autonomy and choice should not dominate the relationship with the child. Rather, we *also* need to embrace the vulnerability that is inherent in close relationships and forego some control over the lives of children, and that is especially the case in non-medical genetic modifications. Relinquishing some of our influence on children's biological lives is part of seeing a child as a gift rather than as a project. Parental authority and planning already extend far into the ways parents influence their children even apart from the use of biotechnologies, but we suggest that genetic enhancements would compromise the relationship to the child too much.

People commonly accept their children as 'gifts', even if they don't talk in terms of a divine 'giver'. They treasure babies irrespective of their health, abilities or beauty. When the philosopher Michael Sandel argued that genetic enhancements would make us view children less as gifts, he was criticized for affirming a religious view of a 'giver' that non-religious people don't share; but that criticism is not persuasive. While some people view the gift as coming from a divine giver, others use it as a helpful metaphor without affirming or denying the involvement of a deity. We should take Sandel's suggestion very seriously that far-reaching genetic modifications 'represent the one-sided triumph of wilfulness over giftedness, of dominion over reverence, of moulding over beholding'.[5]

For Christians, children should be respected and cherished for who they are as persons, as lives that come as gifts from God. As such, they deserve unconditional welcome into the world. It is an uncomfortable fact about human life that many aspects are gifts, not achievements. This should reinforce the belief that all human persons have worth and dignity regardless of what they can or cannot do. Germline modifications would clearly add to the emphasis on performance and achievement when thinking about what a human child should mean to us.

Others have connected the understanding of children as 'gifts' with the distinction between 'begetting' and 'making'. Classic Christian creeds call God the Son 'begotten from the Father', thus emphasizing the close relationship with God the Father. Some ethicists see 'begetting' as a way of characterizing a personal, non-manipulative relationship between parents and children. This is contrasted with technological procedures that are described in the language of 'making' and that are said to reduce children to products of our own clever creating. The natural and non-designed process of 'begetting' children is thus seen as protecting an element of mystery in the child. By editing their child genetically, parents would extend their own ambitions into the next generation. Those advocating such a contrast between begetting and making do not typically speak of the child's human dignity, rights and autonomy, but a connection can be seen between this argument and the modern case for self-determination, which asks not whether procedures are in keeping with *nature* but whether *persons* are used instrumentally for other purposes.[6] This includes the view that when parents pursue their personal ambitions with certain technological interventions, their relationship with the child is compromised. Nevertheless, a strong contrast between begetting and making requires further reflection. At least some aspect of planning plays a legitimate role in human reproduction, and we cannot simply play nature (begetting) off against culture (making), though we should be wary of a manipulative attitude towards persons. With these caveats, this critique of biotechnological manipulation is best directed against non-medical enhancement applications.

A new technology like germline editing brings hopes and fears. Even if only a small percentage of parents decide to modify their child's DNA, whether to reduce risk of a disease or to enhance the child's physical or mental capabilities, this could result in some form of genetic one-upmanship. Other parents could feel obliged to modify their children's DNA in order to level the playing field, and to keep up with the latest genetic fashion. The philosopher Robert Sparrow has even suggested that successive technological advances could lead to a form of genetic obsolescence, in which today's enhancements soon become out of date, when newer and better enhancements become available.[7] This may be similar to the way that computer software has developed, where recent updated versions soon appear inadequate. Genetic enhancement to Life 2.0 may seem inadequate as soon as Life 3.0 becomes available. Children born with today's enhancements may be outdone by children born with later more powerful enhancements, and children born several years later will have even greater enhancements. Would parents regret that they had their child when the latest enhancements were not yet available? Today's enhanced child may be seen as 'yesterday's child' in only a matter of years.

Advocates of germline editing may find these scenarios alarmistic. They may be or they may not be, but we should also question some of the unspoken assumptions. Will parents and children find that germline enhancements have broadly been 'worth the effort', even without potential competition or obsolescence? Will the promised greater capacities lead to a better life? For example, a more powerful memory might, as we have suggested, not improve life but render people unable to let bygones be bygones. The capacity to function well with less sleep may enable greater achievement, but it is not clear that greater achievement will increase life satisfaction.

Increased capabilities and greater choice can sometimes be good. There are of course types of vulnerability and impairment that we should seek to reduce, and we should not romanticize them. However, efforts to create greater autonomy can be unhelpful if we ignore the important social characteristics of our lives that involve interaction with other people. Continual striving for maximum personal choice can make us less ready to accept the claims that other people make on our lives, yet that is precisely what happens in close personal relationships.

People with disabilities, commenting on public life, often say that they do not receive the recognition they deserve, or that they suffer from social exclusion or discrimination. Although society may decide that greater physical functioning makes for better lives, people with disabilities are the ones who are often least likely to benefit from these enhancements, and so their opportunities for social participation decrease, producing even greater exclusion. For people with disabilities, the highest priority is not to hope for new medical advances to rid them of their impairments, and they often don't see their disability as primarily a medical issue. Rather, they are more concerned with social exclusion, discrimination and prejudice. How, then, does a striving for more biotechnological choices impact them?

As a society, of course we do not condone openly hostile behaviour towards people with disabilities, and those arguing for wide-ranging biotechnological enhancements rightly deny that they intend any harm. However, what message would a striving for physical perfection send to such people? What does embryo editing to prevent genetic disease say to those who already suffer from the condition? The philosopher Jonathan Glover, who is in favour of genetic enhancements, suggests that such questions make an important point, 'given the existence of some ugly attitudes towards disabled people'.[8] He also argues that that does not disqualify someone else's attempts to improve life in their own way. Presumably no harm is intended, and some people with disabilities may say. no offense taken. Yet the critical questions about what 'message' is given and heard are not merely about vague sentiments between people. Biomedical technologies can also have a subtle influence on wider circumstances of life, from images of people with perfect bodies that we see in the media to practicalities like the accessibility of cafés and the availability of suitable public transport.

If we idealize the personal right to choose enhanced physical functions, then our everyday behaviour may have consequences for people with disabilities, though this may occur at the level of our subconscious attitudes. In the things that we do and the ways that we plan, many of us do not consider that there will be less able-bodied people around who will need extra consideration. Will a trend towards improvement distract our attention from those who have different aspirations about health and fitness? Our natural inclination is first to consider our own greater autonomy; and the fact that our culture prizes health, vigour and intelligence higher than other moral attributes tends to direct our attention away from those who do not have the same aspirations.

This should not be taken to imply that a desire for improved ability is necessarily bad. Some new treatments of medical conditions may offer a realistic path back to a socially active, creative life. Yet as we seek to strike a good balance, as a society we should give greater consideration to those who are disabled, not just because *they* are created in the image of God but because those who are not disabled are also created in the image of God. For them, social relationships are also an essential part of what it means to be human.

The image of God: the Christlike aspect

The Apostle Paul adds one final twist to the picture of the human person as the image of God, stating that this is a work in progress. Since Jesus Christ is the only person who truly and fully is the image of God (2 Cor. 4.4), the Christian's aim is to become more Christlike and so to manifest the 'image of God' in greater degrees (Rom. 8.29). As Paul describes, in physical and social hardships, Christians sigh in afflictions (2 Cor. 4; Rom. 8). Nevertheless, as Christians they participate in God's power 'in weakness' (2 Cor. 12.9), keeping their treasure in fragile 'clay jars' (2 Cor. 4.7). Paul has the sure hope that Christians are being transformed into the likeness of God, and while that is a long-term process, they already have the dignity of the image of God, whatever their circumstances. Paul reminds his readers of God's Spirit, who is crucial in this transformation. This is the same Spirit who shapes Christians into a community that honours those who are otherwise seen as unworthy (1 Cor. 12.12–31; Rom. 12.1–8). So the two key motifs that we have discussed so far are modified, to yield hope, community and dignity in the midst of distress. Although Paul did not minimize suffering, he viewed it as a situation in which dignity and a different kind of strength are uniquely revealed (2 Cor. 12.9).

Moreover, it is not the perfect whom Christ calls, but those who acknowledge impairments and moral failings. That does not mean that we should deliberately seek after difficult situations that we would rather avoid. Yet even in difficulties or hardships, there is community and hope, and life is still to be affirmed as good and valuable. In this sense there is even a subversive streak in the Bible: in Christ, 'God chose what is foolish in the world to shame the wise; God chose what is weak in the world to shame the strong' (1 Cor. 1.27). Ultimately, our hope does not lie in wisdom and strength. No amount of genetic editing will achieve the perfection of our minds and bodies.

One challenge for this perspective is whether 'strength in weakness' is a realistic possibility. Psychologists have studied how people sometimes claim to feel satisfied under conditions that others would consider to be unpleasant. This casts doubt on the ethical significance of perceived human life satisfaction, a phenomenon they call adaptive preference formation, in which we aspire only to those things that we think are easily within our grasp. This is similar to one of Aesop's fables, where a fox persuaded himself that the fruits that were out of his reach weren't desirable anyway. Similarly, we might convince ourselves that it is not worth striving for any improvements, rather we should be satisfied with our present situation, although it is not really desirable. In part this phenomenon may explain how people maintain a sense of autonomy even when they are afflicted by events that are beyond their control. For example, in some societies men exclude women from formal education. Martha Nussbaum presents as an example of adaptive preference formation how women may sometimes even be content without basic educational opportunities. However, women should still have access to higher education, regardless of the subjective evaluations.

Not only have we appealed to Paul's sense of strength in weakness, but one element in our argument against enhancements has been the case that people with disabilities report high levels of life satisfaction. Are they deceiving themselves into believing that their lives are as good as they possibly can be, and have they instead stoically accepted the status quo, passively assuming that a better life is beyond their reach? If people become content too easily, whatever their life's situation, then maybe we should not take such subjective evaluations too seriously. Sometimes people can be lulled into a false, passive sense of contentment.

Nevertheless, while some people may respond in this way, it is generally not the case for people with disabilities who report high levels of life satisfaction. After all, they often do strive to reduce their exclusion from society, arguing for greater freedom of movement and a reduction in artificial obstacles. Many people with disabilities seek good health care and speak out against things they perceive as unfair in society. People with disabilities who report high levels of life satisfaction are not pretending that their difficulties do not exist, and they are not keeping themselves from acknowledging any suspected unhappiness due to restrictions. If they feel content with their lives nevertheless, we have no reason to doubt them.

Instead, the idea that seemingly 'objective' improvements make us happier in a straightforward way should be questioned, even on an empirical basis. After an increase in social participation, for example, people do report more satisfying lives. However, the law of diminishing returns often applies after the increase has exceeded a certain threshold value. For example, in the area of economics, an increase in earnings does not lead to a proportionate increase in happiness. In a

2010 study in the USA, psychologists reported that an increase in income up to about $75,000 per year improved people's self-reported emotional happiness. However, the trend did not continue beyond that sum: many people who received pay increases above that did not report greater happiness.[9]

It may therefore not be appropriate to seek to use biotechnology to enhance our capacities beyond a given high medical standard. The outcome of aiming for increased life satisfaction through genetic enhancements is doubtful. And the social situation of those with disabilities is likely to deteriorate in a society that endorses enhancements. The Apostle Paul's hope for future redemption and his affirmation of life in the face of weakness should not prompt us to romanticize impairments and rule out uses of biotechnology that could be realistic and genuinely helpful. Yet too strong an emphasis on biotechnological transformation can be in tension with the value of a life that is lived in hope and community, even if an impairment causes discontent.

Conclusion: striking a good balance

What, then, should we make of the different aspects of seeing humans as being made in the image of God? There is some tension between them. The royal dignity attributed to humans speaks for medical efforts in favour of health and vigour. While human dignity should not mean free, untrammelled choice, a certain striving for autonomy is justified from this perspective. Nevertheless, the relational aspect of the image of God implies that our vulnerabilities should not be eliminated across the board, since to be in personal relationships necessarily implies some degree of vulnerability. In keeping with this approach, we have also affirmed that life is a precious gift that is not simply at our disposal. While we cannot deny the legitimacy of some biomedical efforts to improve life, a simplistic maxim that disability is bad and vigour is good is in contrast to the gift, which is worthwhile even in adversity. Paul's perspective on the image of God, finally, suggests that the image was fully revealed in Christ and that Christians strive to be conformed to his image. Religious communities can be subversive as they embody an understanding of strength and wisdom that differs from the usual social conventions and affirms those whom society does not honour.

We suggest that this tension is a productive one, as it challenges us to strike a balance. In Christian theology, two prominent thinkers have embraced similar tensions, which help to do justice to the various factors in particular situations. Karl Barth spoke paradoxically about 'a hastening that waits'. We should never be complacent, but try to make a difference by acting confidently ('hastening'), and yet we should never think that our actions make all the difference, but wait, just as one waits to receive a gift rather than demanding it. Just what this means in real terms will strongly depend on specific factors in different biographical and social situations. Barth was not reflecting on the biotechnological measures at issue here, but we suggest that some biotechnological measures like PGD can be a justified as a 'hastening' in pursuing health. Whether a couple will want to make use of this technology in a particular situation, or whether this reduces the value of life even in adversity, will be their own decision of conscience. By contrast, germline modifications, medical or enhancing, seem to 'hasten' the efforts for physical functioning too much, and diminish the aspect of waiting.

The theologian Dietrich Bonhoeffer, who died as a resistance fighter against the Nazis, also spoke of charting a course between two poles that are in tension with each other: resistance and surrender. He reflected:

I've often wondered here where we are to draw the line between necessary resistance to 'fate' and equally necessary submission. Don Quixote is the symbol of resistance carried to the point of absurdity, even lunacy ... resistance finishes by losing its meaning in reality and is dissipated in theories and fantasies, while Sancho Panza represents complacency, slyly accepting things as they are.[10]

Genetic technologies were not available for some time after Bonhoeffer's death, but the provocative image of Don Quixote may capture something of the quest for enhancements. The 'ingenious gentleman' strives for a worthy goal, pursuing chivalry and patriotic service, yet ends up fighting against windmills. It could seem like another worthy goal to pursue fairness and practise resistance against the arbitrary nature of the 'genetic lottery' that gives particular genetic dispositions to some but not to others. In seeking to correct, even to perfect nature, such resistance rashly seeks to maximize life, seeking vague, shape-shifting possibilities that may never materialize. Quixote contrasts with Panza's resignation that bears all burdens, overly suspicious of 'playing God' even with more realistic, measured biotechnological possibilities. We have described how the various suggestions for genetic enhancements may sound attractive, but upon reflection are unlikely to improve life. Similarly, in two lines of a poem called 'Stations on the Way to Freedom', Bonhoeffer described the realistic action that avoids the phantasies of perfection. It is in this spirit that society should negotiate what human dignity means for contemporary bioethical decisions, charting a course that respects both the autonomy and freedom of the human individual, while acknowledging the dignity that comes from within the vulnerability of personal relationships:

> Doing and daring not what is arbitrary, but the right thing, not being suspended in a world of possibilities, but being undeterred in clasping the real.[11]

Notes

1 Jon Cohen, 'Narrow Path charted for Editing Genes of Human Embryos', *Science* 369, issue 6509 (11 Sept. 2020): 1283.

2 Ruth Macklin, 'Dignity is a Useless Concept', *BMJ* 327, issue 7429 (2003): 1419–20.

3 Hilary Marlow, 'The Human Condition', in *The Hebrew Bible: A Critical Companion*, John Barton, ed. (Princeton, NJ and Oxford: Princeton University Press, 2016), pp. 293–312.

4 Martha C. Nussbaum, *Frontiers of Justice: Disability, Nationality, Species Membership* (Cambridge, MA: Belknap Press, 2006), ch. 3.

5 Michael Sandel, *The Case Against Perfection: Ethics in the Age of Genetic Engineering* (Cambridge, MA: Belknap Press, 2007), p. 85.

6 While Michael Sandel does not argue with the distinction between 'begotten' and 'made', he sees the gift-character of life best preserved in naturalness; thus observers have noted with surprise that Jürgen Habermas wrote the preface to the German edition of his *The Case against Perfection*. Sandel, *Plädoyer gegen die Perfektion: Ethik im Zeitalter der genetischen Technik*, 2nd edn (Berlin: Berlin University Press, 2008), pp. 7–14.

7 Robert Sparrow, 'Yesterday's Child: How Gene Editing for Enhancement will Produce Obsolescence – and Why it Matters', *The American Journal of Bioethics* 19, issue 7 (2019): 6–15.

8 Jonathan Glover, *Choosing Children: The Ethical Dilemmas of Genetic Intervention* (Oxford: Oxford University Press, 2006), p. 33.

9 Daniel Kahneman, *Thinking, Fast and Slow* (London: Penguin, 2012), p. 396.

10 Dietrich Bonhoeffer, *Letters and Papers from Prison* (Dietrich Bonhoeffer Works, Vol. 8) (Minneapolis, MN: Augsburg Fortress, 2010), pp. 303–4 (trans. modified A. Massmann).

11 Bonhoeffer, 'Stations on the Way to Freedom', in *Letters and Papers from Prison*, Vol. 8, pp. 512–13 (trans. modified A. Massmann).

Further reading

Science

Denis Alexander, *Genes, Determinism and God* (Cambridge: Cambridge University Press, 2017).

Francoise Baylis, *Altered Inheritance: CRISPR and the Ethics of Human Genome Editing* (Cambridge, MA: Harvard University Press, 2019).

Nessa Carey, *Hacking the Code of Life: How Gene Editing will Rewrite our Futures* (London: Icon, 2019).

Jennifer Doudna and Samuel Sternberg, *A Crack in Creation: The New Power to Control Evolution* (London: Vintage, 2018).

Jim Kozubek, *Modern Prometheus: Editing the Human Genome with CRISPR-Cas-9*, 2nd edn (Cambridge: Cambridge University Press, 2018).

Jamie Metzl, *Hacking Darwin: Genetic Engineering and the Future of Humanity* (Naperville, IL: Sourcebooks, 2019).

McKusick-Nathans Institute of Genetic Medicine, *OMIM: Online Mendelian Inheritance in Man: An Online Catalog of Human Genes and Genetic Disorders*, www.omim.org/ (accessed 24.9.2020).

John Parrington, *Redesigning Life: How Genome Editing will Transform the World* (Oxford: Oxford University Press, 2016).

Robert Plomin, *Blueprint: How DNA Makes us Who we Are* (London: Penguin, 2019).

Theology and theological ethics

Denis Alexander, 'Healing, Enhancement and the Human Future', *Cambridge Papers* 28 no. 1 (2019), https://tinyurl.com/yxm7oqts (accessed 5.10.2020).

Conference of European Churches, *Moral and Ethical Issues in Human Genome Editing: A Statement of the CEC Bioethics Thematic Reference Group* (Strasbourg: 2019), https://tinyurl.com/y53soxvd (accessed 5.10.2020).

Ellen Charry, 'God and the Art of Happiness: The 2007 Alfred Palmer

Lecture', video recording, 2 Dec. 2007, Seattle Pacific University, https://tinyurl.com/yylqn2vw (accessed 29.9.2020).

Gerald McKenny, *Biotechnology, Human Nature, and Christian Ethics* (Cambridge: Cambridge University Press, 2018).

Neil Messer, 'The Ethics of Cognitive Enhancement: A Theological Perspective', video recording, 26 May 2015, The Faraday Institute for Science and Religion (Cambridge, UK), https://tinyurl.com/y6epdz7l (accessed 29.9.2020).

Nancey C. Murphy and Christopher C. Knight, eds, *Human Identity at the Intersection of Science, Technology and Religion* (Farnham: Ashgate, 2010).

Annette Weissenrieder and Gregor Etzelmüller, 'Illness and Healing in Christian Traditions', in *Religion and Illness*, A. Weissenrieder and G. Etzelmüller, eds (Eugene, OR: Wipf & Stock, 2016), pp. 263–305.

Michael Welker, *Creation and Reality* (Minneapolis, MN: Fortress Press, 1999).

Further resources

Havi Carel, *Illness: The Cry of the Flesh*, 3rd edn (London: Routledge, 2018).

Jonathan Glover, *Choosing Children: Genes, Disability, and Design* (Oxford: Oxford University Press, 2006).

Sheila Jasanoff, *The Ethics of Invention: Technology and the Human Future* (New York: Norton, 2016).

David Kahneman, *Thinking, Fast and Slow* (London: Penguin, 2012).

Philippa Levine, *Eugenics: A Very Short Introduction* (Oxford: Oxford University Press, 2017).

National Academy of Sciences, *Heritable Human Genome Editing* (Washington, DC: The National Academies Press, 2020), https://doi.org/10.17226/25665 (accessed 5.10.2020).

Nuffield Council on Bioethics, *Genome Editing and Human Reproduction: Social and Ethical Issues* (London: 2018), https://tinyurl.com/yy6pjgne (accessed 5.10.2020).

Tom Shakespeare, *Disability: The Basics* (Abingdon: Routledge, 2018).

Index of Names and Subjects